Contents

Foreword: Laws of Form: Spencer-Brown at Esalen, 1973
   Louis H. Kauffman 5

Articles

Gurus in the Mud
   Cliff Barney 7

AUM Conference Transcripts: Session One: Monday Morning March 19, 1973
   Cliff Barney and Kurt von Meier 17
AUM Conference Transcripts: Session Two: Monday Afternoon March 19, 1973
   Cliff Barney and Kurt von Meier 35
AUM Conference Transcripts: Session Three: Tuesday Morning March 20, 1973
   Cliff Barney and Kurt von Meier 55
AUM Conference Transcripts: Session Four: Tuesday Afternoon March 20, 1973
   Cliff Barney and Kurt von Meier 64

Commentary: The First Message From Space
   Kurt von Meier and Cliff Barney 77

The Flagg Resolution Revisited
   James M. Flagg and Louis H. Kauffman 87

Paper Computers, Imaginary Values and the Emergence of Fermions
   Louis H. Kauffman 107

Regular Features

Column

Virtual Logic—Who Shaved the Barber?
   Louis H. Kauffman 163

ASC Column

The Destabilizing Cybernetics of Implausibility: The Anti-Anthropocentric Crisis
   Zane Gillespie 169

The Artist for this issue is Paul Snelson, II. Full color art at www.chkjournal.com. For more information on this artist see https://www.paulsnelson.com.

Cover Art

Snelson, P., II. (2019). *Cybernetic Art Matrix Revitalized* (detail). Computer graphic.

CYBERNETICS & HUMAN KNOWING
A Journal of Second-Order Cybernetics, Autopoiesis & Cyber-Semiotics
ISSN: 0907-0877

Cybernetics and Human Knowing is a quarterly international multi- and trans-disciplinary journal focusing on second-order cybernetics and cybersemiotic approaches.

The journal is devoted to the new understandings of the self-organizing processes of information in human knowing that have arisen through the cybernetics of cybernetics, or second order cybernetics its relation and relevance to other interdisciplinary approaches such as C.S. Peirce's semiotics. This new development within the area of knowledge-directed processes is a non-disciplinary approach. Through the concept of self-reference it explores: cognition, communication and languaging in all of its manifestations; our understanding of organization and information in human, artificial and natural systems; and our understanding of understanding within the natural and social sciences, humanities, information and library science, and in social practices like design, education, organization, teaching, therapy, art, management and politics. Because of the interdisciplinary character articles are written in such a way that people from other domains can understand them. Articles from practitioners will be accepted in a special section. All articles are peer-reviewed.

Subscription Information

Price: Individual £75; Institutional: £166+VAT (online); £203 (online & print). 50% discount on full set of back volumes. Payment by cheque in £UK (pay Imprint Academic) to PO Box 200, Exeter EX5 5HY, UK; Visa/Mastercard/Amex.
email: sandra@imprint.co.uks

Editor in Chief: Søren Brier, Professor in semiotics at the Department of International Culture and Communication Studies attached to the Centre for Language, Cognition, and Mentality, Copenhagen Business School, Dalgas Have 15, DK-2000 Frederiksberg, Denmark, Tel: +45 38153246. sb.ikk@cbs.dk

Editor: Jeanette Bopry, Instructional Sciences, Ret.
jeanette.bopry@gmail.com

Special topic editor: Dirk Baecker, Zeppelin University,
dirk.baecker@uni-wh.de

Associate editor: Dr. Paul Cobley, Reader in Communications, London Metropolitan University, 31 Jewry Street, London EC3N 2EY. p.cobley@londonmet.ac.uk

Managing editor: Phillip Guddemi, The Bateson Idea Group, Sacramento CA 95812, USA. pguddemi@well.com

Joint art and website editor: Claudia Jacques
cj@claudiajacques.org

Editorial Board

M.C. Bateson
George Mason Univ.
Fairfax VA 22030, USA

Dirk Baecker
Zeppelin University, D-88045
Friedrichshafen, Germany

Rafael Capurro
Dept. of Information Science,
Hochschule der Medien, Stuttgart
University, Germany

Bruce Clarke
Dept. of English, Texas Tech
University, Lubbock, TX 79409,
USA

Marcel Danesi
Semiotics and Communication
Studies, Toronto U. Canada

Ranulph Glanville
CybernEthics Research
Southsea, UK

Ernst von Glasersfeld
Amherst, Mass., USA

Jesper Hoffmeyer
Dept. of Biological Chemistry
Univ. of Copenhagen, Denmark

Michael C. Jackson
The Business School, University of
Hull, UK

Louis Kauffman
Dept. of Math. Stat. Comp. Sci.
Univ. of Illinois, Chicago, USA

Klaus Krippendorff
School of Communications
University of Pennsylvania, USA

George E. Lasker
School of Computer Science
Univ. of Windsor, Canada

Ervin Laszlo
The General Evolution Group
Montescudaio, Italy

Humberto Maturana
Univ. de Chile, Santiago, Chile

John Mingers
Kent Business School,
Univ. of Kent, UK

Edgar Morin
Centre d'Etudes Transdisciplinaires
Sociologie, Anthropologie, Histoire
(CETSAH), Paris, France

Winfried Nöth
Wiss. Zent. f. Kulturforschung
University of Kassel, Germany

Roland Posner
Arbeitsstelle für Semiotik
Technische Universität, Berlin

Bernard Scott
Academician of the Int. Academy
of Systems and Cybernetic Sci.

Fred Steier
Interdisciplinary Studies University
of South Florida

Ole Thyssen
Dept. of Management, Politics and
Philosophy, Copenhagen Business
School, Denmark

Robert Vallée
Directeur Général, Org. of Systems
and Cybernetics, Paris, France

C&HK is indexed/abstracted in *Cabell's Journal* and *PsycInfo*
Journal homepage: www.chkjournal.com
Full text: www.ingenta.com/journals/browse/imp

Consulting editors :
Niels Åkerstrøm Andersen,
Copenhagen Business School
Hanne Albrechtsen, Royal School
of Librarianship, Copenhagen
Argyris Arnellos, University of
the Aegean, Syros, Greece
Dan Bar-On, Ben-Gurion Univ.
of the Negev, Beer-Sheva, Israel
Jon Frode Blichfeldt, Work
Research Inst., Oslo, Norway
Geoffrey C. Bowker
Santa Clara University, USA
Philippe Caillé, Inst. of Applied
Systemic Thinking, Oslo, Norway
Sara Cannizzaro,
London Metropolitan University
Paul Cobley, Lanauge and Media,
Middlesex University, UK
Guilia Colaizzi
U. Of València, Spain
Finn Collin, Philosophy, U. of
Copenhagen
John Collier, Philosophy, U. of
Natal, Durban, South Afrika
Allan Combs
CIIS, San Francisco, CA, USA
David J. Depew, Dpt. Communic.
Studies, U. of Iowa, USA
Anne Marie Dinesen
Univ. of Aarhus, Denmark
Daniel Dubois, Inst. de Math. U.
de Liege, Liege, Belgium
Per Durst-Andersen, Copenhagen
Business School, Denmark
Charbel Niño El-Hani
Federal Univ. Bahia, Brasil
J.L. Elohim, Instituto Politecnico,
Nacional Mexico City, Mexico
Claus Emmeche, Niels Bohr Inst.
Copenhagen, Denmark
Donald Favareau, National
University of Singapore
Christian Fuchs, ICT&S Center,
U. of Salzburg, Austria
Hugh Gash, St. Patrick's College,
Dublin 9, Ireland
Carlos Vidales Gonzales,
U. of Guadalajara, Mexico
Christiane Herre, Xi'an Jiaotong-
Liverpool U., Suzhou, China
Steen Hildebrandt, The Aarhus
School of Business, Denmark
Wolfgang Hoffkirchner, ICT&S
Center, U. of Salzburg, Austria
Stig C Holnberg, Mid-Sweden U.
Seiichi Imoto, Philosophy,
Hokkaido U. Sapporo, Japan
Ray Ison, Centre for Complexity
and Change, Open Univ., UK
Kathrine E.L. Johansson,
Copenhagen Business School
Pere Julià, Inst. f. Advanced
Stud. C.S.I.C., Palma de Mallorca
Dr. Shoshana Keiny, Education,
Ben-Gurion U. Negev, Israel
Ole Fogh Kirkeby, Copenhagen
Business School, Denmark
Kalevi Kull, Dept. of Semiotics
Tartu University, Estonia
Marie Larochelle, Dpt. de Psycho-
pedagogic, U. of Laval, Canada
Allena Leonard, Viable Systems
International, Toronto, Canada

Floyd Merrell, Purdue Univ.,
West Lafayette, IN, USA
Gerald Midgley,
University of Hull
Asghar Minai, Sch. Architecture,
Howard U., Washington DC, USA
Jean-Louis Le Moigne, France
Vessela Misheva, Uppsala
University, Sweden
Andrea Moloney-Schara
Georgetown Family Center
Arlington, Virginia, USA
Ole Nedergaard,
Copenhagen Business School
Massimo Negrotti
Univ. Degli Studi Di Urbino
IMES, Urbino, Italy
Per Nørgaard, Royal Academy of
Music, Aarhus, Denmark
Makiko Okuyama, Ohmiya Child
Health Center, Japan
Nina Ort, Inst. für deutsche
Philologie, Ludwig-Maximilians
University, München
Marcelo Pakmann
Behavioral Health Network,
Springfield, MA, USA
Charles Pearson, Austell,USA
Andrew Pickering, Dept. of
Sociology, University of Exeter
Bernhard Poerksen, Tubingen
Univ., Germany
Peter Pruzan, Copenhagen
Business School, Denmark
Lars Qvortrup, Dean of School of
Education, University of Århus
Axel Randrup, Center for Inter-
disciplinary Studies, Roskilde
Yveline Rey, Centre d'Etudes et
de Recherches sur l'Approche
Systémique, Grenoble, France
Robin Robertson, Editor,
Psychological Perspectives, LA
Steffen Roth, ESC Rennes School
of Business, France
Wolff-Michael Roth, SNSC,
Univ. of Victoria, Victoria, BC
Stan N. Salthe
Natural Systems, New York
Eric Schwarz,
U. De Neuchâtel, Schweiz
Inna Semetsky, IASH, Univ.
Newcastle, NSW, Australia
Erkki Sevänen, University of
Joensuu, Finland
Göran Sonesson, Lunds
Universitet, Sweden
Bent Sørensen
Aalborg U., Denmark
Stuart Sovatsky
California Inst. Integral Studies
Torkild Thellefsen, Dept. of
Communication, Aalborg U.
Ole Nedergaard Thomsen,
Copenhagen Business School
Robert E. Ulanowicz, Chesapeake
Biological Lab., USA
Mihaela Ulieru
University of Calgary, Canada
Bruce H. Weber, Dept.Chemistry,
California State University
Maurice Yolles, Management
Systems, John Moores U., UK
Gerard de Zeuw,
Lincoln University, UK

Copyright: It is a condition of acceptance by the editor of a typescript for publication that the publisher automatically acquires the English language copyright of the typescript throughout the world, and that translations explicitly mention *Cybernetics & Human Knowing* as original source.

Book Reviews: Publishers are invited to submit books for review to the Editor.

Instructions to Authors: To facilitate editorial work and to enhance the uniformity of presentation, authors are requested to send a file of the paper to the Editor on e-mail. If the paper is accepted after refereeing then to prepare the contribution in accordance with the stylesheet information at www.chkjournal.org

Manuscripts will not be returned except for editorial reasons. The language of publication is English. The following information should be provided on the first page: the title, the author's name and full address, a title not exceeding 40 characters including spaces and a summary/ abstract in English not exceeding 200 words. Please use italics for emphasis, quotations, etc. Email to: sbr.lpf@cbs.dk

Drawings. Drawings, graphs, figures and tables must be reproducible originals. They should be presented on separate sheets. Authors will be charged if illustrations have to be re-drawn.

Style. CHK has selected the style of the APA (*Publication Manual of the American Psychological Association*, 5th edition) because this style is commonly used by social scientists, cognitive scientists, and educators. The APA website contains information about the correct citation of electronic sources. The APA Publication Manual is available from booksellers. The Editors reserve the right to correct, or to have corrected, non-native English prose, but the authors should not expect this service. The journal has adopted U.S.English usage as its norm (this does not apply to other native users of English). For full APA style informations see: apastyle.apa.org

Accepted WP systems:
MS Word and rtf.

Snelson, P., II. (2009). *What's the Score?* Computer graphic.

Foreword: Laws of Form
Spencer-Brown at Esalen, 1973

Louis H. Kauffman

This special issue of *Cybernetics and Human Knowing* contains rare material related to G. Spencer-Brown's book *Laws of Form* and its contents.

In 1973 there was a conference at Big Sur at which G. Spencer-Brown discussed his calculus with a group of scientists. This was the AUM Conference at Esalen, and the scientists consisted in an assortment of remarkable individuals exploring the cutting edge of human consciousness and culture, including Alan Watts, Ram Dass, John Lilly, Heinz von Foerster, Kurt von Meier and others. One of the participants, Cliff Barney has written about this conference and has long been a keeper of the transcripts of Spencer-Brown's talks and the concurrent discussions.

In addition to Cliff Barney's introduction to the AUM conference, "Gurus in the Mud," we here publish the conference transcripts and an article, "First Message From Space," by Cliff Barney and Kurt von Meier reflecting on the conference. The transcripts are a remarkable amalgam of the thinking of G. Spencer-Brown and the questions and comments of the participants in AUM. The transcripts carry the same lucidity that infuses *Laws of Form*. Other than reading *Laws of Form* itself, I know of no better introduction to it than these lectures of G. Spencer-Brown.

The other articles in this issue are of equal interest.

The paper on the Flagg Resolution by James Flagg and Louis Kauffman is a careful exposition of a solution to self-contradiction based on non-locality in the space of the text of the contradictory element. The Flagg Resolution states that if $A = \sim A$ then this equality can undergo substitution in a formula only if this substitution is done for every occurrence of A. Non-locality and time are closely related and the paper handles this theme elegantly and with regard to the points of view of G. Spencer-Brown in chapter 11 of *Laws of Form*.

The paper "Paper Computers and the Emergence of Fermions" by Louis Kauffman is an expansion of an unpublished document "Paper Computers" by Louis Kauffman in 1980. The present paper explains how, by allowing Boolean algebra to become circular in a diagrammatic form, one constructs not only paradoxical or oscillating elements, but memories—structures that have two stable states, and all the necessary ingredients to construct counting, arithmetic operations and the basis for computers. This can be viewed as a discussion of the underpinnings of chapter 11 of *Laws of Form*, but this paper is self-contained. The paper then goes on to discuss the concept of imaginary values that arises from the cybernetics of Boolean circularity and it shows how algebras related to physics, Fermion algebra and Clifford algebra arise naturally from the temporality of elementary oscillations. The paper ends by showing how the Dirac equation in quantum mechanics unfolds in this context.

Regular features in this issue include two columns. The Virtual Logic column by Louis Kauffman is a new take on the Barber paradox and the Russell paradox, based on satire, mirrors and the key observation of Douglas Harding that no person can (in the absence of mirrors) perceive his or her own head. The world is in the eye of the beholder, but that eye does not exist (stand forth) for the beholder.

The ASC Column by Zane Gillespie is about the structure of implausibility (a plurality of things, interconnected, individually innocuous, but together a challenge to human complacency) in music, art and cybernetics. Here is a formulation that goes beyond simple paradox to address revolution in a new way.

Artist of the issue is Paul L. Snelson, II; he is a visual artist, scholar, and writer working on cybernetic arts and as a producer and business management software quality analyst at Thryv. He lives in Dallas, Texas, where he completed an MFA in Arts and Technology and an MA in Visual Art and Cybernetics at the University of Texas at Dallas.

Currently, Snelson is constructing a documentary hybrid 2D/VR film based on the all-encompassing design and structural influence of, Roy Ascott's Cybernetic Art Matrix, virtually as a three-dimensional graph. The film will showcase the function of grid structure in a myriad of exemplary designs including science, technology, architecture, visual arts, music, and theatre. Ascott's prescient concept of The Age of the Cybernetic Art Matrix / The C.A.M. era, is now the current point of view including an omni-media perspective now actualized in virtual reality and all standard viewing formats. Snelson is also working on an opera entitled *Modus Operandi*, which features a constructed theatrical set as the C.A.M., a three-dimensional cubic scaffolding structure where the fourth dimension of time as sound and music inhabits both space and time in a live black box theatrical production. The libretto text is available at this link: https://www.paulsnelson.com/modus-operandi.

We are happy to be able to present this very special issue on laws of form in *Cybernetics and Human Knowing*, and we particularly wish to thank Søren Brier for all the help and support he has given us over the many years of discussion and writing about these concepts.

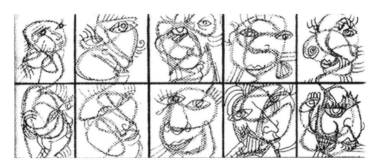

Snelson, P., II. (2009). *Capturing Character Series* (detail). Ink on canvas.

Gurus in the Mud

Cliff Barney[1]

Despite its grand name, and its spectacular setting at the Esalen hot baths in a cliff by the sea, the AUM[2] conference on the laws of form proved something of a bust for many of the participants, who failed to achieve intellectual enlightenment in the week it lasted. Nor was it a success for the intellectual focus of the show, British mathematician G. Spencer-Brown, who fell out with the sponsors and left after spending only two days of the scheduled seven. Without Spencer-Brown, the others organized tentative seminars around various ways to ask "What do you suppose he meant?" They didn't have a lot of success.

Laws of form, the calculus, and G. Spencer-Brown, who put it into the marked state, had enjoyed an ambiguous reputation since the book, *Laws of Form*, was published in the U.S. in 1972. Spencer-Brown himself was an enigma: On one side, he was a student of Ludwig Wittgenstein and Bertrand Russell, the logico/mathematical heavyweights of the early 20th century, and Russell had praised the calculus. On the other, Spencer-Brown was notoriously eccentric and had published, as well as mathematics, poetry and belles lettres in which he hinted at deeper meaning to his mathematics than mere descriptions of functionality. One epigraph to *Laws of Form* is from William Blake: "Tho obscured, this is the form of the Angelic land." This is not simply a pun; in the text that follows, Spencer-Brown promises to show laws that operate where the concepts of mathematics and religion have degenerated and are not distinguished.

For reasons that will be suggested below, the Esalen meeting left most of its participants more puzzled at the end than when they had assembled. The laws of form have remained almost an intellectual curiosity, largely ignored by the mainstream and investigated mostly on the fringe, both mathematical and mystical. For a time, the book was out of print in the United States.

At present, however, a Google search on *laws of form* returns over a million hits (of which the first five, and probably many others, concern strictly legal matters). Over the years, new editions of the book have been published: a paperback version of the first American edition from the Julian Press, a second American edition from a Portland publisher, Cognizer Co., and later a German translation, *Formenkalkul Gesetze der Form* from Bohmeier Verlag.

One computer scientist, William Bricken of the University of Washington, has interpreted laws of form as a computer language (LOSP, written in Lisp) and has built the language into a chip to power a very fast video board for virtual reality applications. In 1994, *EE Times*, a trade magazine, printed several articles on the use

---

1. Email: cbarney@jeffnet.org
2. AUM is an acronym for American University of Masters, created by the always inventive Alan Watts.

of the laws of form in designing logic circuits (their original application). Several computer scientists, including the inventor of the computer language called Forth, Chuck Moore, were at that time designing logic circuits that use imaginary values in calculating states, thus simplifying chip design, as demonstrated in the *Laws of Form*. The scientists and engineers are thus plugging onward, but the mystics and metaphysicians, not understanding math, have mostly given it up (with the exception of a laws of form community, composed at least partly of mathematicians, organized around Spencer-Brown himself in England). This is too bad because *Laws of Form* is too rich to be left to logic and technology. Therefore perhaps it is time, 45 years after the AUM conference, to stir this pot again.

Kurt von Meier and I were among the twenty or so people present that March week at Esalen to meet and talk with Spencer-Brown. It was a transforming experience. I had no idea what I was getting into, beyond the prospect of spending a few pleasant days on the California coast, steaming in the Esalen baths and basking in the intellectual glow of the hip literati of the day—Alan Watts, John Lilly, Ram Dass (formerly Richard Alpert, who was Tim Leary's partner in taking LSD public in the 1960s),[3] Heinz von Foerster, Karl Pribram, Michael Murphy, Charles Tart, Stewart Brand, John Brockman, various other scientists, artists, psychologists willing at least to speculate that there might be some formal relationship between hard science and math and the more spiritual pursuits that were then beginning to become popular. At that time, Lilly was seriously trying to talk with dolphins and had not yet announced that he was a visitor from another star system. Watts, an Anglican clergyman, had become a cult figure in the hippie world with his lively, entertaining books on the psychedelic experience and Buddhism. (Even today, more than forty years after his death, Watts's taped lectures are still occasionally broadcast.) Brand's *Whole Earth Catalog* had enchanted a whole new demography with its stunning array of concise reviews and pointers to offbeat products and intellectual resources that were considered useful for living independent, productive lives.

Among the entries in the catalog was a review of *Laws of Form* signed by Prof. von Foerster. And what a review it was! "At last the *Laws of Form* have been written," it began, surely one of the most arresting sentences possible, implying compactly that laws of form existed and that at least one person had been eagerly awaiting their publication. *Laws of Form*, von Foerster (1970, p. 14) wrote, was "a 20th-Century transistorized version of Occam's razor."

The review was accompanied by a brief excerpt from the book—terse, even cryptic:
- Draw a distinction.
- Call it the first distinction.
- Call the space in which it is drawn the space severed or cloven by the distinction.

---

3. Ram Dass, who had by then become a hero of the human potential movement, took a brief break from the conference to visit the then imprisoned Leary. When he returned, he reported "Tim told me 'You get better and better and I get worse and worse.'"

- Call the parts of the space shaped by the severance or cleft the sides of the distinction or, alternatively, the spaces, states, or contents distinguished by the distinction.
- Let any mark, token, or sign be taken in any way with or with regard to the distinction as a signal ... .

Draw a distinction? What did that mean? Why did it matter what we called it? What was Spencer-Brown getting at? That was the point of the AUM conference, since a lot of people, none of them principally mathematicians, were beginning to read into the idea of distinction something basic about the way that human beings organized information and in fact constructed their entire reality. So if one could in some way codify the idea of distinction, hmm, well, perhaps that might be worth looking into, especially if one were a psychologist or a therapist or an anthropologist or any kind of religious, artistic, or social analyst looking for a firm scientific peg on which to hang one's aura. Lilly and Watts prevailed on George Gallagher, a Hawaiian psychiatrist, to put up some money, rented part of Esalen for a week, hired Spencer-Brown as a lecturer, and sent out invitations to their friends and colleagues to come and hear him, gratis. It was an invitation not to be turned down.

I obtained entrance to this illustrious group only through my association with Kurt von Meier, then a professor of art and mythology at California State University, Sacramento, and an old friend of John and Toni Lilly's. I had just moved in with Kurt and his family at the quaint Diamond Sufi ranch in the Napa Valley, and we were casting about for something to do together. The AUM conference looked like just the ticket. Kurt was a natural for the list; a true polymath, he combined a thorough grounding in his academic discipline, art history, with a wide-ranging interest in all of the arts and sciences, and a willingness actually to learn and practice other disciplines, rather than simply to read and write about them. Kurt showed his classes a thousand ways in which all human culture is one, in lectures delivered with the precision and detail of a Joseph Campbell and frequently in the style of the late Lord Buckley.

I had no such claim to entry, but had long been trained in the art of free-loading by the profession that knows it best, journalism. I called a friend who was an editor and wangled an assignment to cover the AUM conference for the *Saturday Review*, a venerable magazine then publishing its final feeble editions from San Francisco. The editors cooed over the guest list, chortled over the site (a hot springs where people of both sexes took off their clothes and sat naked together in the baths), and gave the piece a working title: "Gurus in the Mud." (They must have confused it somehow with the mud baths in Calistoga north of San Francisco.)

So it was that we rolled up our magic carpet and piled into my Volkswagen bus one Sunday and rolled down the valley toward Esalen—not just Kurt and I, but his lady friend Mary Evans as well, with their baby Amanita, and our friend Paula Reineking and her then boyfriend Chuck, who were also living at the ranch. The others hadn't exactly been invited, but in those days we went everywhere together.

In 1973, Esalen was at the beginnings of its popularity as the center for innovative therapeutic and psychological activity. Its hot baths, steaming out of a Big Sur cliff high over the Pacific ocean, had been made the centerpiece of a bustling commerce in spirituality and health. Fritz Perls had done much creative work on the practice of Gestalt therapy while in residence there. Pioneering research on psychedelic drugs had taken place in Esalen's friendly environs. The practice of communal nude bathing hinted at sexual adventure. Doing a weekend at Esalen soon became fashionable for many folks searching for new answers to existential angst. So many came that the Esalen management invested in a motel a few miles up the road from its main center to handle the overflow. It was here, in the South Coast motel, that the AUM conference took place.

Spencer-Brown showed up, British to the nines, and plunged gamely into a discussion of the laws of form; but from the very start, the audience's unfamiliarity with mathematics caused the discussion to flounder. Spencer-Brown tried to make clear the difference between mathematics, a calculus, and the interpretation of that calculus; but the conferees had trouble with even this simple distinction. We got tangled up in the idea of the mathematics of a state of mind—how could there be such a thing?

Even more subversive to our goal of understanding, however, was the difference of opinion, unknown to us at the time, between Spencer-Brown and the sponsors, particularly John Lilly. Apparently it concerned money, but whatever the substance, it resulted in Spencer-Brown packing up after two days and heading back to England, leaving the rest of us with the rest of the week to tease out some of the clues he had left.

Lilly and Watts organized a series of seminars in which individuals could give their own interpretations of the laws of form, in the hope that from this rough collaboration could come some sort of consensus. This was not, alas, to be, although some of the presentations, particularly that of Heinz von Foerster (Wittgensetein's nephew and himself a logician of note) managed to shed light on the process of calculating without numbers. Kurt, who from the very first understood Spencer-Brown's method, if not his matter, gave a version of the Buddha's Flower Sermon—he silently covered a blackboard with symbols of the calculus, wrote at the bottom "Homage to all teachers," and bowed himself out. A rump underground, led by Stewart Brand, washed its hands of the whole affair on grounds that Spencer-Brown himself had let us all down, and a few people actually went home, foregoing the remaining days of Esalen hospitality.

I was as lost as the rest until Heinz, at the end of his lecture, illustrated the formal nature of Spencer-Brown's calculus by singing us a couple of the mathematical expressions. This is not as silly as it may sound. Written music is, after all, simply an agreed-on notation in which musical intervals are represented by steps on a scale. All Heinz did was map certain notations in the calculus into musical scales, assign values to the notes, and read the music.

Later he and I discussed this idea privately, and even managed to harmonize on some of the expressions in the calculus; and from these exercises I got my first inkling of what Spencer-Brown meant by laws of form, not laws of thought or idea or physics or anything else, simply form—and how it was represented—and of how powerful a tool it was to isolate purely formal values.

While still enchanted by the idea of setting the calculus to music, I providentially encountered a group of musicians who were staying at Esalen. They picked up on the process immediately and had no trouble whatever understanding the process of making music from mathematics. We found rehearsal space and began to compose; and later, as the Brown Cross Chorale (two flutes, dulcimer, viola, temple bell, and vox humana) performed our initial (and only) work, based on Consequence 1 of the *Laws of Form*,[4] at the closing session of the conference, Sunday morning on the deck at the main Esalen lodge.

Despite this minor triumph, however, the mood on that occasion was anything but clear. It had been, we all agreed, a most interesting time, quite stimulating, a great deal of fun—but what had we really learned? No one was willing to commit. The prevailing emotion was relief at not having to think about laws of form any more.

Not for Kurt and me, though. By now thoroughly imbued with the magic of the laws, and convinced that Spencer-Brown, wittingly or unwittingly, was a major teacher and discoverer, on the order of a Dante or a Newton, we went back the Napa Valley and spent the next three years studying his texts and puzzling out some of the hints he had provided. We had the *Laws of Form*, of course, plus two other Spencer-Brown books: *Probability and Scientific Inference*, an earlier publication, and *Only Two Can Play This Game*, a book of poetry and short pieces that he had published under the cabalistic name of James Keys. (It is the latter book that got Spencer-Brown into his deepest trouble with his mathematical colleagues, since in it he specifically relates religious forms such as angels with mathematical theorems.) Kurt had made tapes of the Esalen sessions, and I transcribed them; these transcriptions are the basis of the present publication.

We began to look at formal structures, dice, and probabilities with new eyes. From these we were led to labyrinths, and from the labyrinth to the Morris dance, and the Cretan double axe, and its relation to the labyrinth; and from Crete to Dedalus, and thence to James Joyce, and gradually we began to see some of the larger patterns, and the laws of form began to make sense.

We never learned how calculate with them. Painstakingly, I had puzzled out the process of pattern recognition by which the laws of form carry out their transformations; having done so, I could see that one would need innate skill and intensive training to become adept in their use. I have never seen this aspect of the laws of form discussed. Transformations in the laws of form are made by substituting one pattern for another, the patterns being shown by the calculus to be equal in value.

---

4. C1, Reflexion, $\overline{\overline{a}} = a$, *Laws of Form*, p. 28, specifically the derivation of this consequence summarized on p. 31. The performance was recorded, but the tape, lamentably, has been lost.

These patterns may be of any scope and may change from step to step in a demonstration, or even within a given step. It is not always obvious—in fact it is seldom obvious—where to look to see pattern similarities.[5]

Nevertheless, one could still invoke the power of the calculus simply by following Spencer-Brown's clear instructions and seeing how it is structured, how it inevitably grows and develops out of the initial instruction quoted above, "Draw a distinction." The difference between this kind of injunctive language and the descriptive language that we use most of the time took on ever greater significance.

As a student and teacher of Gestalt therapy, I was familiar with Fritz Perls's remark that injunctions, commands, constitute the clearest form of communication; however, I still considered the indicative mood as the main vehicle of speech and the imperative as an appendage, as it is in common speech and language teaching. Spencer-Brown showed that instructions must maintain an inner coherence that is not demanded of descriptions, and pointed out that it is not a matter of opinion what the result of carrying out an instruction will be, whereas descriptions constantly give rise to differences of opinion. That is, if I give you a musical score, which may be interpreted as a series of instructions, and say "Play this music," it is not a matter of opinion what the music will sound like. (Whether the music is pleasant or not may definitely be a matter of opinion.) After many years of living with this distinction I find myself increasingly uninterested in anyone's opinion about anything.

The first thing I did on returning home was to write, with Kurt's felicitous assistance, the piece for *Saturday Review*. Given their vision of the various Gurus rolling around in mud, the editors did not look favorably on my account of turning the calculus into some form of New Age music. Before they could reject it completely, however, the *Saturday Review* itself folded, and I was left with an orphan manuscript. I sold it for $100 to my friend Don Stanley, who was then editing a lively Marin County weekly called the *Pacific Sun*, and in due course it saw print, plugged on page one and with marvelous illustrations by a staff artist, in one of which Spencer-Brown's visage replaces the Charioteer on the seventh Tarot key, with images of Lilly and Watts adorning the animals who draw the chariot.[6] The piece drew two letters, both of which cast aspersions on my intelligence, if not my basic sanity; still, I was happy enough that it was published.

Kurt, meanwhile, began teaching laws of form to his art history classes at Sacramento State. Together we devised a syllabus and put together a reading list that embraced texts as far apart as the *I Ching*, Robert Graves's *The White Goddess* and D'Arcy Thompson's classic *On Growth and Form*, with the transcript of the Esalen conference as a centerpiece. "Learn to draw distinctions," Kurt told his astonished students, who were accustomed to being told "Repeat after me... ."[7]

---

5. The method is spelled out by Spencer-Brown in chapter 6 of *Laws of Form*, p 28.
6. This article is available online at https://www.kurtvonmeier.com/who-is-g-spencer-brown-and-where-is-that-marvelous-music-coming-from
7. A description of an examination for one of these courses may be found on Larry Barnett's Kurt von Meier website at https://www.kurtvonmeier.com/art-113c-midterm-exam.

And together, we began composition of "The Ömasters," our fantastic fictive account of how the laws of form could be promulgated to the world. We adopted Karl Pribram's suggestion to use the form of a Sufi teaching story, and Kurt began to make up characters and situations, mingling Tarot cards and current headlines into an utterly riotous adventure that combined space opera, a parody of American democracy, and an East-West fantasy farce. We would haul our typewriters out on the deck and write alternate pages, exchanging them and cackling with glee as we read aloud to whoever would stay around long enough to listen. There weren't many of those; and although Kurt and I had a lot of fun, we wound up with stacks of manuscripts and tapes and a reputation as possibly loony or possibly onto something really big, with no possibility of anyone else ever knowing which it was.

We had one flirtation with mainstream publishing,[8] when John Brockman induced an editor from Dutton to make a special trip to the ranch to consult with us; but although he laughed more heartily than most at our pages, he could not, to put it kindly, see their commercial possibilities. Stewart Brand agreed to consider publishing the Esalen transcript in the *Whole Earth Catalog*, and then managed to lose the manuscript, an event that led to some coolness on both sides.

After three years, the Napa Valley community began to break up and I found myself needing to go back to civilization and start earning a living. It was time to move on from the laws of form. We made a bundle of all of our manuscripts and mailed it off to Spencer-Brown at his last known address in Cambridge, together with a letter saying how much we had enjoyed writing them and that we hoped he liked them too.

A few weeks later he phoned us from England. We were the only people at Esalen who'd got it, he said. He'd love to come to the States and visit us and how would we like to get together on a new publishing venture? He would offer as capital some fifty thousand copies of his latest book (Parsons, 1975), a dictionary of music in which tunes were classified according the up or down sequence of their musical intervals. (I have never actually seen this book.) He also sent along a wonderful manuscript, a group of fables called "Stories Children Won't Like" (never published, so far as I know), and announced that he had discovered a proof for the four-color theorem using the laws of form.[9]

---

8. Actually, we had two: I have only recently seen for the first time a most encouraging letter to John Brockman from an editor at Doubleday; it's online at https://www.kurtvonmeier.com/omasters-rejection-letter. Kurt never showed me the letter; I suspect because he hated to revise, as it suggests we do.
9. At this time two American mathematicians, Kenneth Appel and Wolfgang Haken, had announced a proof of this classic theorem (which states that four colors are sufficient to color a map on a surface of genus 0). Their proof used a computer to exhaustively test a finite but large collection of configurations, at least one of which must appear in any possible map. Spencer-Brown took a different approach and tried to prove that any map could be generated by four colors. A book by Spencer-Brown on the four-color theorem was announced by Scribner's in 1974, but it apparently never saw print. The most recent edition of *Laws of Form* contains accounts of Spencer-Brown's work on the four color problem and his work on the Riemann Hypothesis and other problems in number theory.
This edition is from Bohmeier Verlag in Leipzig, Germany; but it is in fact in English, the fifth such edition, and includes also a new introduction by Spencer-Brown, in which he tells in comic detail how he induced Bertrand Russell to endorse his work.

Of course Kurt and I immediately put ourselves at Spencer-Brown's disposal. We had been working with no outside encouragement for three years, and here the fellow calls up and says yep, that's it, let's play ball. We knew we were right, understand. We had already shrugged off the feedback, from a well-known biographer of Wittgenstein, that we were mathematical innocents whose interpretation would be laughed at by Spencer-Brown. Still, to have direct contact with the source was stimulating, to say the least.

The story of what happened when Spencer-Brown—or James Keys, his alter ego, as we came to know him personally—came to live at the Diamond Sufi ranch in Oakville is one of the best in the whole saga, funny and embarrassing and strangely bittersweet. It deserves telling if only for the truly mortifying events surrounding the reception we gave Spencer-Brown on Alan Watts's houseboat in Sausalito. Watts had by then died, but he left behind a society that continued to promote his tapes and books; this group very kindly lent us the houseboat and invited haute Marin County for wine and cheese on a Sunday afternoon, to chat with the visiting master. James never showed up, having found a popsy and driven her down the coast for a Big Sur weekend, and I found myself making apologies to a crowd that was only slightly more puzzled by not seeing the guest of honor than they had been for being invited to meet him in the first place. Spencer-Brown and the lady showed up ten days later and took me to lunch at the Cliff House in San Francisco, and no one said a mumbling word about the Alan Watts Society.

We had many other adventures of a similar order, some of them uproariously funny, and we never got anywhere in promoting Spencer-Brown's literary venture. In fact I never found out what it was. We did promote a couple of Spencer-Brown seminars, at Nepenthe near Esalen and at Wilbur Hot Springs in Colusa County, and then we went separate ways. James moved to Palo Alto, where he taught for a while at Stanford University and at the Xerox Palo Alto Research Center. The last time I saw him he was living in a house with no furniture, and he took twenty dollars from me at Frisbee golf.

Enjoyable as it was, our experience with Spencer-Brown himself was only a coda to the sessions Kurt and I held in Oakville during the writing of "The Ömasters." The fact that we are not accomplished writers of fiction (consider how few people are) does not affect our vision of the laws of form as a powerful paradigm. And Spencer-Brown himself remains a figure worth study. I am unqualified to comment professionally on the laws of form as mathematics; I have, however, worked for several decades as a writer and editor, during which time I learned something about prose construction, and Spencer-Brown is a master of it. Whatever his merits as a mathematician, he writes elegantly and clearly. He has complete command of the forms of English, using words with great precision and recalling for us their derivation.

Some of this mastery spills over into his speech, and the Esalen transcript is a remarkable document. That it can still find an audience so many years after its creation is remarkable in itself. Spencer-Brown disavowed it when Dirk Baecker asked his permission to have a German translation prepared; still, it deserves close reading. Witty and subtly funny, it is explicitly descriptive of some of the connections Spencer-Brown makes between the language of mathematics and the language of experience.

The first session is particularly valuable for his description of the origin of laws of form, and how it simplified switching logic so as to make the equipment that performed it cheaper and yet more reliable. Despite the mathematical ignorance of most of his audience, Spencer-Brown struggles mightily to explain just now math is constructed, and the relation of what he has done to the rest of mathematical literature. He goes to great lengths to show how imaginary Boolean values are related to imaginary numerical values, and in the process shows the irrelevance of Russell's and Alfred North Whitehead's theory of types that underlies their much-unread attempt to derive mathematical principles from logic, *Principia Mathematica*.[10]

Session Two devotes space to the difference between algebra and arithmetic, a distinction fundamental to laws of form, which Brown says is the non-numerical arithmetic to non-numerical Boolean algebra. The session also includes entertaining excursions on proofs and demonstrations and prime numbers.

Session Three contains a deep discussion of Spencer-Brown's conception of form and distinction, the nature of his token of the marked state, the cross, and the requirements for communication between not-speaking people. And the final session outlines Spencer-Brown's conception of what he calls the *five eternal levels*, (a kind of mathematical analog of Heaven) and the generation of Time. Throughout there are entertaining anecdotes, attention to etymologies, and mini-lectures on how one performs mathematics.

Though old and even partially disowned, the Esalen texts retain *baraka*. They have enjoyed a kind of underground life on the Internet, popping up occasionally using material copied from an original laws of form site maintained by Interval Research computer scientist Dick Shoup before his death (where the Esalen transcripts were first published from our transcript). This version is the only print publication and will reward its readers with new insight into Spencer-Brown's work.

References

Parsons, D. (1975). *The directory of tunes and musical themes*. Cambridge, UK: Spencer-Brown & Co.
Spencer-Brown, G. (1972). *Laws of form*. New York: Julian Press.
Von Foerster, H. (1970). Laws of form. In S. Brand (Ed.), *Whole Earth Catalog*, #1090, Spring, p. 14. Menlo Park, CA: Portola Institute.

---

10. Spencer-Brown has said that when he approached Russell with proof that the theory of types was unnecessary, Russell was relieved, since he himself believed that the theory was arbitrary and unnecessary.

George Spencer-Brown (from the dust jacket of Keys, 1972)

AUM Conference: Session One
Monday Morning, March 19, 1973

transcript created and edited by Cliff Barney
from recordings by Kurt von Meier

JOHN LILLY: G. Spencer-Brown—enigmatic figure to say the least. His book preceded him. We know less about him than we know about Carlos Casteneda.[1] His book expresses a good deal that is impersonal, universal; and hence, the man is kind of hidden by the book. I began to find the man when I found his second book, published under a pseudonym, James Keys, called *Only Two Can Play This Game*, and as soon as I read footnote one, I suddenly realized what *Laws of Form* was all about. And with that I will leave the discussion to G. Spencer-Brown: James.

G. SPENCER-BROWN: Well, I don't know what to say. It is a great pleasure, a great honor, to be here. I don't feel that I deserve the honor in any way. It is also—I think I feel rather nervous, as this audience has so many and so different qualifications.

I don't hope to do anything but really answer any questions that anybody has to ask about the nature of what I was trying to do when I began to write *Laws of Form* and the very different answer, what I actually found, that had appeared when I had finished the book, which was not what I had set out to do. I guess that is the only way that I can begin to talk about the work, which is as far as I am concerned entirely impersonal. It has as little to do with me personally as anything I can imagine. I have no particular attachment to it. I wouldn't do it again if anybody asked me to. I was conned into writing it by thinking that it would have an entirely different effect from what it did have; and, in completing it, I unlearned what I learned, the kind of values that present-day civilization inculcates into us soon after we are born. And I learned that it was all the same anyway, whichever state one went into. It is only by assuming that some states, or that a state, one state, may be better than another, that the universe comes into being. The universe, as I then discovered, is simply the result of if it could be that some state had a different value from some other state. But that is to start at the end.

Mathematics and Logic

SPENCER-BROWN: At the beginning, what I was concerned to do was—having left the academic world and gone to living in London, I became an engineer. And I was detailed to make circuits for the use of the new transistor elements that were coming

---

1. At the time, Carlos Casteneda, a UCLA anthropologist, was at the height of his popularity for his books about Don Juan, a Yaqui shaman who, Casteneda reported, had trained him as a shamanic apprentice.

into being for making special purpose computers. I was employed by a firm then known as Mullard Equipment, Ltd., a branch of the Phillips organization, and I was employed not because of any engineering qualifications but because I had taught logic at Oxford and it was recognized that, in fact, the study of logic in some form or another was necessary to designing circuits involving on-off switches. So I began with the very specific task of applying what I knew to these circuits, to see if we could devise rules for designing that would save money.

I rapidly found that the logic I had learned at the University and the logic I had taught at Oxford as a member of the logic faculty wasn't nearly sufficient to provide the answers required. The logic questions in university degree papers were childishly easy compared with the questions I had to answer, and answer rightly, in engineering. We had to devise machinery which not only involved translation into logic sentences with as many as two hundred variables and a thousand logical constants—ANDs, ORs, IMPLIES, et cetera—not only had to do this, but also had to do them in a way that would be as simple as possible to make them economically possible to construct—and furthermore, since in many cases lives depended upon our getting it right, we had to be sure that we did get it right.

For example, one machine that my brother and I constructed, the first machine I mentioned in *Laws of Form*, counts by the use of what was then unknown in switching logic; it counts using imaginary values in the switching system. My brother and I didn't know what they were at the time, because they had never been used. We didn't at that time equate them with the imaginary values in numerical algebra. We know now that's what they are. But we were absolutely certain that they worked and were reliable, because we could see how they worked. However, we didn't tell our superiors that we were using something that was not in any theory and had no theoretical justification whatever, because we knew that if we did, it would not be accepted, and we should have to construct something more expensive and less reliable. So we simply said, "Here it is, it works, it's OK," and British Railways bought it, we patented it, and the first use for it was for counting wagon wheels. It had to count backwards and forwards, and we had one at each end of every tunnel. When a train goes into a tunnel, the wagon wheels are counted, and when it comes out, they are counted. If the count doesn't match, an alarm goes out, and no one is allowed in that tunnel—at least, not very fast.

This had to be not only a very reliable counter, it had to count forwards and backwards, because…you know what happens when you get on the train: It goes along and then it stops and then it goes backwards for a bit, goes forwards. So, if the train was having its wheels counted, and then, for any reason, ran out of steam and got stuck and then slipped back, then the counter had to go backwards. So all this we had—but we made it in a way which was very much simpler than, and very much more reliable because of being so simple, than the counters in use at that time, which amounted to much more equipment, many more parts. This device was patented. The patent agent of the British Railways, who patented it—of course, we never told him what he was writing out. We just told him to write this down. And it worked, it has been used ever since, and though there have been many disasters in British Railways since that time, not a one of them has consisted of any train running into a detached wagon in a tunnel. Fingers crossed, touch wood.

We made many other devices, and during this time I realized that, unfortunately, it would be necessary for somebody to write up the mathematical basis of the new principles that we were using. And I realized that if I didn't do this, it would be very hard to find anybody who would. And so I started writing it up.

After we had been using the new principles for about a year, most of the discoveries had been made. I wasn't quite sure of the theoretical basis of some of them. For example, to realize that what we were using in the tunnel was imaginary Boolean values to get a perfectly safe, reliable answer, which was quite definite. This I didn't realize for another six years. But most of the principles, by that time, I did realize. They were the whole of the mathematical basis of what we were using, which was switching algebra, commonly called Boolean algebra and the algebra of logic.

I must point out for emphasis at this time that the switching use and the use in checking a logical argument are two entirely different applications from the same mathematics. The same mathematics underlie both, but it is not the same as any one of its interpretations. In other words, the mathematics in *Laws of Form* is not logic, logic is one of many of its interpretations. Just as, when one does electronics, the electrical application is not itself the mathematics but one of the interpretations of mathematics.

Boolean Mathematics

SPENCER-BROWN: Logic, in other words, is itself not mathematics, it is an interpretation of a particular branch of mathematics, which is the most important non-numerical branch of mathematics. There are other non-numerical branches of mathematics. Mathematics is not exclusively about number. Mathematics is, in fact, about space and relationships. A number comes into mathematics only as a measure of space and/or relationships. And the earliest mathematics is not about number. The most fundamental relationships in mathematics, the most fundamental laws of mathematics, are not numerical. Boolean mathematics is prior to numerical mathematics. Numerical mathematics can be constructed out of Boolean mathematics as a special discipline. Boolean mathematics is more important, using the word in its original sense: what is important is what is imported. The most important is, therefore, the inner, what is most inside. Because that is imported farther. Boolean mathematics is more important than numerical mathematics simply in the technical sense of the word *important*. It is inner, prior to, numerical mathematics—it is deeper.

Now at the beginning of 1961, the end of 1960, having set out, first of all, as an exercise in what I thought was logic, I began to write it out. Realized it wouldn't fit. Took it back. Took it back, got it farther and farther back until I got it right back, what we had been working on in engineering and the principles of it, right back to the simplest ground and the simplest obvious statements about the ground one had constructed. And at the end of 1960, I had become conscious that the whole of this mathematical world could be taken to the simplest of all grounds, and the grounds were only that one drew a distinction. The defining of a distinction was a separation of one state from another—that is all that was needed.

This was all that was needed to make the whole of the construction which is detailed in *Laws of Form*, and which will suffice for all the switching algebra, train routing, open/shut conditions, decision theory, the feedback arrangements, self-organizing systems, automation and, amusingly enough, the logic by which we argue, the logic that is the basis of the certainty of mathematical theorems. In other words, the forms of argument which are agreed to be valid in the proof of a theorem in mathematics. To give you a simple one: "If x implies not-x, then not-x." That is a commonly used argument—I can be sure that it is valid by the principles of the mathematics itself that underlies it.

The arguments used to validate the theorems in *Laws of Form*, as we now begin to see, are themselves validated by the calculus dependent upon those theorems. And yet, in no way is the argument a begging of the question. Now this is rather hard to understand, and perhaps it may come up in discussions later. *Petitio principii*, begging the question, is not a valid argument; it is a common fallacy. In no way is the question begged, but in producing a system, in making its later parts come true, we use them to validate the earlier parts; and so the system actually comes from nothing and pulls itself up by its own bootstraps, and there it all is.

Nowhere does this become more evident than in this first and most primitive system of non-numerical mathematics; and I am quite sure—no, I will not say I am quite sure, when one says "I am quite sure" it means one is not quite sure—and I guess, I guess that why it is a branch of mathematics so neglected hitherto is that it is a bit too real. It is a bit too evident what game one is playing when one plays the game of mathematics.

If one starts much further away from the center, then you don't see the connections of what you are doing. You don't see that what comes out depends on what you put in. You can devise an academic system that goes on the assumption that there is objective knowledge, which we are busy finding out. We have come along here with wide-open eyes, and what we see over there—we come along and we give a demonstration, and we write it out, et cetera, and when somebody says, "But just what is it that gives the formula that shape? Why is it that shape and not some other shape? What is it that makes these things true? What is it that makes it so that when we see this, what makes it so—why isn't it otherwise?" And the stock answer is: "Ah, well, that is how it is, and that is the mystery."

Mystery, after all, doesn't mean that we scratch our heads and look in astonishment and amazement. Mystery means something closed in. A mystic, if there is such a person, is not a person to whom everything is mysterious. He is a person to whom everything is perfectly plain. It's quite obvious. And the person who designates himself a non-mystic, and has nothing to do with that kind of woolly thinking, is a person, an ordinary academic, who writes down his mathematical formulae, and when people say "Why do they look like that, why don't they look some way else?" "Well, they just are that way; it's perfectly justified by mathematics. If you do mathematics, that's what you have to learn to do." In fact, when one starts from the beginning, there is nothing to learn. There is everything to unlearn, but nothing to learn.

KURT VON MEIER: When you told us about tunnels I saw the great psycho-cosmic projection of images and tales of the parable of Plato's cave. So I imagine you have provided us with the parable of the tunnel. It is in the shape of the hole of doughnut, topologically, so we could look for the seven-color rainbow with which to color it. See the map of a torus; it is seven colors—

SPENCER-BROWN: Do you know the proof of that?

VON MEIER: I think there have been many attempts—

SPENCER-BROWN: It has been proved. I haven't actually followed the proof of that, although the question is interesting topologically.

Coloring a Torus

SPENCER-BROWN: I believe the principle by which you can prove that you can color the surface of a torus with seven colors is wholly different from the principle by which—if it is true that you can color the surface of a sphere with four colors—by which it would be proved. I have a feeling about the general question—have you looked at it like this: in any surface, if you take a small enough part of it, you have again the problem of the plane. Because a small enough part of any surface is, for all intents and purposes, a plane. You take a little bit of a torus and now you have a four-color theorem again. As long as you don't go round…and round. The four-color theorem is contained in every theorem about surfaces… And so it is a different kind of theorem.

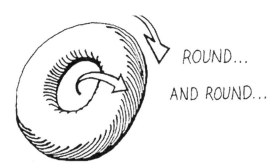

Drawing adapted from Spencer-Brown's illustration on a whiteboard—CB

MAN: Is there a question of which is prior, or that sort of thing?

SPENCER-BROWN: Well, I think there is a difference like this: You can prove the color number—like, a torus is seven. It needs a minimum of seven colors to be sure of coloring a map on a torus is that no two bordering areas are the same color. I think that these are all decidable using what is currently allowed in mathematics: Boolean equations of the first degree only. I think that why we cannot, why we never have decided the four-color theorem and a number of other theorems, like Fermat's last, and Goldbach's, is not that they are undecidable. The questions can be asked, and, Wittgenstein was right about this, if a question can be asked, it can be answered. There is a definite answer to all these questions, these theorems are actually true or false. Why we cannot decide them is that they need, in fact, at least equations of the second degree in the Boolean argument, and possibly use of the imaginary values. The answer would be quite definite, Just as the answer to the number of wagon wheels, although the actual logic has used imaginary values. The answer is quite definite.

VON MEIER: We are getting into interesting space. I can see a wagon wheel as something of an iron doughnut, if you like. It's been put on the axle of a train. And if you see the tunnel of the British railway system and consider the space that flows through that tunnel as going around the whole earth and inside the tunnel and then around up to the sky, what we have, in fact, is a super-distended doughnut. And what you were explaining about the wagon wheels passing through the doughnuts, then, would seem to me to be something to do with the space that's ruled by the spirit of inside the doughnut, provided by the doughnut hole. What kind of changes can take place inside that domain? It is in a field—there are analogs in physics that define the inside of the doughnut as continuous, and get, nevertheless, in a special way distinguished from the space of the rest, the outside, of the doughnut.

SPENCER-BROWN: I pass on that one.

ALAN WATTS: A human being is topologically a doughnut.

Unlearning

LILLY: Have you formulated or recommended an order of unlearning?

SPENCER-BROWN: I can't remember having done so. I think that, having considered the question, the order of unlearning is different for each person, because what we unlearn first is what we learned last. I guess that's the order of unlearning. If you dig too deep too soon you will have a catastrophe; because if you unlearn something really important, in the sense of deeply imported in you, without first unlearning the more superficial importation, then you undermine the whole structure of your personality, which will collapse.

Therefore, you proceed by stages, the last learned is the first unlearned, and this way you could proceed safely. Related to what is in the books, we know they say that in order to proceed into the Kingdom, one must first purify oneself. This is the same advice, because the Kingdom is deep. What we talk of in the way of purification is the superficial muck that has been thrown at us. First of all that must be taken off, and the superficial layers of the personality must be purified. If we go to the Kingdom too soon, without having taken off the superficial layers and reconstructed in a simpler way, then there is a collapse. The advice is entirely practical. It is not a prohibition. There is no heavenly law to say that you may not enter the Kingdom of Heaven without first purifying yourself. However, if you do, the consequences may be disastrous for you as a person.

This is why in psychological, in psycho-therapeutic treatment, normally the defenses are strong enough. As the psychiatrists will usually tell you, "If I push in this direction, you will be able to withstand me if you really need to." And it is much the same in all medicine. A rule I learned—I guess one learns it here, John—in the treatment of physiotherapy, manipulation of the limbs, et cetera, we are allowed to go and pull them around with our little strength, but not to use machinery, because that may break something. The body can normally defend against one other body, and you don't usually break anything as long as you use one physiological equipment against one other. Usually the same; one mind against one other, the other mind is strong enough to withstand it. Start using other methods, drugs and/or mechanical treatment, and there you may do damage. You may get past defenses which were there in order that the personality should not be broken down too much, too soon.

WATTS: There is a value assumption in here about what is broken down. What is disaster, what does that mean?

SPENCER-BROWN: Well, it is a value judgment, true enough. In reality, it is all the same. In reality, it is a matter of indifference, but we are not here in reality. We are here on a system of assumptions, and we are all busy maintaining them. On that system, then we can say, "Well, that will keep the ship afloat, and this will pull the plug out and we will all sink."

Degree of Equations and the Theory of Types

DOUGLAS KELLEY: As we go from second-order equations to third-order, I imagine you would like to maintain your two, and only two states, the marked and the unmarked. And if that is the case, in going from second to third order, do you get a more generalized concept of time, or a little different—I am just wondering what a third order equation would look like.

SPENCER-BROWN: Well, I think you mean degree equations; first-, second-, and third-degree equations.

KELLEY: A degree of indeterminacy, yes.

SPENCER-BROWN: Now, basically, once we have gone into the second degree, and it applies in numerical mathematics as elsewhere, you have added another dimension to your system. In going to higher degrees, you don't really add; because, when you start with the form, the form is just having drawn a distinction. You now have two states. It's the simplest, widest term I could possibly use here—states on earth, anything, you see, states of mind, anything at all. You have two states, which are distinguished. That is all you need.

Now, the whole of the first-degree equation in the Boolean form are in terms of these two states. When you do this peculiar thing of making something self-referential that is making the answer go back into the expression out of which the answer comes, you now automatically produce this set of possibilities which are well-known in numerical mathematics and of which everyone's been terrified of looking at in Boolean mathematics. And Russell/Whitehead were so frightened of these, that they just had a rule with no justification whatsoever that we just don't allow it, we don't even allow people to think about this.

Now what nobody saw was that in numerical mathematics we had this going for years. As I showed in the preface to the American edition to *Laws of Form*, any second-degree equation—perhaps, for those of you who don't know it, perhaps I should put it up on the blackboard if nobody has objections to my using chalk.

You see, what Whitehead/Russell didn't allow, was a self-referential statement; they didn't allow to say that this statement is true. *["This statement is false" written on board.]* Suppose that this statement is true, then it can't be true because it says that it is false. O. K. then, supposing it is false, then it must be true because it says that it is false. And this is so awful, so terrifying, that they said, "Right. We will produce a rule. We call it the theory of types to give it a grand name." The theory of types says—it is as much unlike what it says as possible, so that when someone says, "Well, what is the rule by which you can't have this?"…"It's the theory of types," so that the people who are learning think that there is a huge theory, you see, and when you understand this theory you will realize why it is that you can't have such a thing. There is no such theory at all. It is just the name given to the rule that anything like this you must do this to. *[Erases "This statement is false."]* That is the theory of types.

What they hadn't done was scratched out something like this. I'll just put it in the mathematical form. *[$x^2 + 1 = 0.$]* You see, Russell, as a senior wrangler, or second wrangler, in mathematics, should have been familiar with this equation. But he never connected it with what he had done.

Now here is an equation which admittedly had a bad name for years. But it was so useful that all of phase theory in electricity depends on it. So let's fiddle with it. Here is our equation. We want to find the roots. We want to find the possible values of *x*. So let's fiddle with this and have a look for them. Well, here we go. Here we just subtract one from both sides; now we'll divide both sides by *x*. Well, *x*-squared divided by *x* is *x*, equals minus one over *x*. *[Writes $x = -1/x$.]* Well, now, we see that we have in fact a self-referential equation. Everybody can see that. Let's have a look at this equation $x = -1/x$, and see whether it is amenable to any form of treatment, psychiatric or something. You have to psychoanalyze it.

The thing that makes the former statement so worrying, so frightening, is that we have the assumption that the statement, if it means anything at all, is either true or false. Here, we have the assumption that the number system runs "-3, -2, -1, zero, 1, 2, 3,"and it goes on infinitely in an exact mirror both ways. So we assume that the number is not zero—zero is meaningless in the logic form. The statement is not meaningless. It is either positive or negative. We have got to make that analogy here. We equate positive with true, and negative with false—it doesn't matter which is which.

So here is our number system as defined. Here is our equation from which we are supposed to find the possible values that x can take. Now, we know that the equation must balance...so first of all we'll seek the absolute numerical value of x, irrespective of the sign, whether it's positive or negative. Now suppose x were greater than one—not bothering about the sign for the moment—suppose it were greater than one, then this clearly would be—not bothering about the sign—less than one. If x were less than one, then you have got something bigger over something smaller, this would be greater than one. So the only point at which it is going to balance numerically is if x is a form of unity. Because you can see perfectly well that if this is greater then that would be smaller, if that is smaller, this would be greater.

So we have only got two forms of unity—plus one, minus one. So we'll try each in turn. So suppose x equals plus one, now we'll substitute for x in this equation and we have minus one over plus one equals minus one. *[Writes: 1= -1/ 1 = -1.]* So you've got plus one equals minus one. So try the other one, there is only one more: x equals minus one. Now we have minus one equals minus one over minus one equals plus one. *[Writes: -1 = -1/-1 = +1.]* So we have exactly the same paradox this time. Instead of *true* and *false*, we have got *plus* and *minus*. So using the theory of types consistently, the whole of the mathematics of equations of degree greater than one must be thrown out. But we know perfectly well that we can use this mathematics. What we do here is that effectively we have an oscillatory system—just as in the case of laws of form, if you put it mathematically, we have $x = \overline{x\rceil} = \overline{\sqcup}$ or a cross going back into itself.

Supposing it is the marked state, then it puts the marked state back into itself, and the marked state within a cross produces the unmarked state outside. So this rubs itself out and so you get the unmarked state fed back in, and so out comes the marked state again.

Well, you see here the paradox which was overlooked by Russell, who wasn't a mathematician, although he was senior wrangler, and by Whitehead, who was, although he wasn't. Well, Russell was a mathematician, he wasn't a man of mathematics. Whitehead was a man of mathematics. Russell knew the forms, but he actually had no instinctual ability in mathematics. Whitehead actually had. But Russell, being a stronger character, was able to program Whitehead, and you will see this if you examine the last mathematical work Whitehead wrote, which is called *A*

Treatise on Universal Algebra with Applications, Vol. 1. I asked Russell where Vol.—I said I had never been able to get Vol. 2, and Russell said, "Oh, he never wrote it."So it's all sort of a mystery. But the mathematical principles of algebra, in the usual complicated way, are set out, including the Boolean algebras, in this volume produced in 1898, an only edition. By that time Russell, who was the stronger of the two characters, had got together with Whitehead to do *Principia Mathematica*, which nobody was ever going to digest—It was a very ostentatious title, because they had chosen the title which Newton had used for his greatest work.

Incidentally, it is an extraordinary thing in the academic world—people are very silent about these things—but it was a very, very presumptuous title, I think, to take for this work. *[Inaudible comment, to the effect, "Hasn't* Laws of Form *been used?"]* Oh, no, nobody has used that title before—no, sir. If I had called it *Laws of Thought*, that was used, many people have used that title, but it was not laws of thought. Oh, no, you are on the wrong track, sir. I am not being presumptuous in taking that title…I have called the book what it is, I have not done what Russell/Whitehead did and taken a very great book and called it by the same title. That is totally different.

Now, this is what they overlooked in the formulation of the theory of types, which simply says you mustn't do this. However, both Russell and Whitehead had done it to get their wranglerships, get their degrees. But they had not done the simple thing of reducing this equation to this to see exactly what it was.

In fact, if you go to the Boolean forms and use something like this—there's your output—and you take it back in, input there and these are transistors used in a particular way, you have what is called a memory. And, if you put *minus* instead of *plus* there, $x^2 - 1 = 0$ instead of $x^2 + 1 = 0$, now what we have here, back in this form here, is our equation. Now we'll put it all in brackets and we'll take out the answer. Now we have exactly the same thing. And just as this is a memory circuit, if this is the marked state here, that must be the unmarked state.[2] And if this is unmarked state, we've got no marked state here, so this will be marked. And we have a marked state feeding itself back into there, and if you rub that out and this goes unmarked, you still have marked here, so it remembers. Equally, if you now put a marked state here, that must be unmarked, and then you can take that off and it doesn't matter because now since you have got unmarked and unmarked this becomes marked, and this, you remember, is unmarked. Similarly here, if you put *plus one* for *x*, you get plus one over plus one equals plus one, there is no paradox. You can also find a different answer for *x*, and that is *minus one*. You get plus one over minus one equals minus one, so that's all right too. So you have, in effect, a memory circuit, and if you put it this way, you can see that. You have an equation with two roots, and this is similarly an equation with two roots. Whatever root you get out, you put back in, and it remembers itself. If you are getting out "plus one," it feeds plus one in there, and it remembers it's plus one. You have a thing to knock it off and turn it into minus one; it

---

2. Transmission of Spencer-Brown's marks on the blackboard has been lost due to the absence of note-taking at the time. The general discussion concerns re-entry at an odd level and at an even level. If odd, as in $\overline{x|}$, we get marked state in and unmarked out, an oscillation. If even, $\overline{\overline{x|}}$, we get marked in, marked out, a memory.

feeds a minus one into there and out comes minus one, here, and it remembers it's minus one. Any equations of the second degree that are not paradoxical—that go through two stages and not one—are the same, and this is a way of producing a memory circuit electronically. It is exactly analogous to this memory circuit numerically. And where, in fact, you put it back, instead of here, you put it back through an odd number such as one, now you have a paradoxical circuit. Because whatever it gets out it feeds back in and it changes. And if you turn that into minus, $x^2 = -1$, you now have a paradoxical equation. It can't remember, it just flutters.

Suppose it were an electric bell; in fact, here under our very noses all the time we have in mathematics the mathematics of the electric bell. And to show how the human mind works, in all the mathematics textbooks it says there is no mathematics of the electric bell. Here it is, all before our eyes. The simplest and most obvious things are the last and hardest to find because we have to get so awfully complicated before we get there.

Feedback and Fluttering

MAN: Are those analogous to positive and negative feedback?

SPENCER-BROWN: Yes. It's all straight feedback. A positive feedback remembers itself, a negative feedback oscillates. We have got the mathematics of the oscillator.

How often do you use this operator *[writes: $i(\sqrt{-1})$]*, which is derived from the paradoxical equation? Now, why is *i* used so much? Because *i* is the state that flutters, is the oscillation. This has been totally overlooked in mathematics, that *i* is in an oscillatory state. Because in order to get over this paradox of *x*-squared equals minus one, we see that we can't use any ordinary form of unity so we invent in mathematics another form of unity and we call it *i*, which is the root that satisfies that equation. And the root that satisfies that equation is that you have plus one, minus one, and here's a state between; and the root that satisfies that equation, whatever it is, it isn't. And this is why *i* is so useful in dealing with that kind of curve—because it is, by its very nature, that kind of curve. *i* is an oscillation.

Time and Space

SPENCER-BROWN: It is really an oscillation defining time; but it is the first time, and, therefore, being the first time, the oscillations are without duration, so the wave has no shape at all. Just as the space of the first distinction has no size, no shape, no quality other than being a state. This is one of the things that tend to upset people. It is part of the mathematical discipline that *what is not allowed is forbidden*.[3] That is to say, what you don't introduce, you can't use. And until you have introduced shape, size, duration, whatever, distance, you can't use it.

---

3. First Canon. Convention of Intention. *Laws of Form*, p. 3.

In the beginning of *Laws of Form*, we defined states without any concept of distance, size, shape—only of difference. Therefore the states in the laws of form have no size, shape, anything else. They are neither close together nor far apart, like the heavenly states. There is just no quality of that kind that has been introduced. It's not needed.

The same with the first time. The first time is measured by an oscillation between states. The first state, or space, is measured by a distinction between states. There is no space for a distinction to be made in. If a distinction could be made, then it would create a space. That is why it appears in a distinct world that there is space.

Space is only an appearance. It is what would be if there could be a distinction.

Similarly, when we get eventually to the creation of time, time is what there would be if there could be an oscillation between states. Even in the latest physics, a thing is no more than its measure. A space is how it is measured; similarly, time is how it is measured. The measure of time is change. The only change we can produce—when we have only two states—the only change we can produce is the crossing from one to another. If we produce an expression, like the ordinary expressions in the algebra, we have to make the crossing. We have to do something about it. We have to operate from the outside. If we produce that cross that feeds into itself, now we don't have to do anything. It is a clock, just as an ordinary distinction is a rule. A rule makes or defines space, and a clock defines time. In making our first distinction all that we have done is introduce the idea of distinction. We have introduced nothing else. No idea of size, shape, distance, and so on. They do not exist, not here. They can be constructed, and they will be, but not yet. They are what happens when you feed the concept back into itself enough times.

Again, when you first construct time, all that you are defining is a state that, if it is one state, it is another. Just like a clock, if it is tick, therefore it is tock. But this time is the most primitive of all times, because the intervals are neither short nor long; they have no duration, just as these states have no size.

There were some books written about time by a man called J. W. Dunne that I read when I was a schoolboy. I realized that he must be right. I also was sufficiently aware of the social context to go along with the general opinion that he was off his head. He wasn't. He was dead right. Time is a seriality, and he was quite right. In order to get a time such as the one we experience, you have to put it back on itself, because in our time you have duration, which you can measure; and you can only measure the duration with another time. In the first time, you have no time in which to measure how long your duration is, and so, naturally, you can't have any duration. Time is something you have to feed back into itself several times. Like the space of this room, where you can actually measure it—you have to have space to measure space.

LILLY: Is that frequency of oscillation either zero or infinity?

SPENCER-BROWN: It is neither. No, it has no duration at all. Just as you can't specify the size of the states of the first distinction.

LILLY: So that it has no determined frequency.

SPENCER-BROWN: No. It can't be infinite, it can't be zero. So, the space determined by the first distinction is of no size.

HEINZ VON FOERSTER: It's just flippety and not frequency.

SPENCER-BROWN: Yes, just flippety.

MAN: And that's saying it could be any size you want.

SPENCER-BROWN: No—you see, all this is a children's guide to the reality, "as if it had some size." It is not right to say it could be any size you want. Because you have to learn to think without size. Anything like that is misleading, just as it's misleading to say this can be any duration you want. It doesn't have duration. It just don't have it. Just like the void don't have quality.

GREGORY BATESON: What about the *then* of logic? If two triangles have three sides, et cetera, then so-and-so. The *then* is devoid of time.

SPENCER-BROWN: Yes. There is no time in logic, because there can't be time without a self-referential equation, and by the rule of types, which is now in operation in the defining of current logic, there is no feedback allowed. Therefore all equations in logic are timeless.

BATESON: So we add sequence without adding duration.

SPENCER-BROWN: If you make a feedback, which Russell and Whitehead disallowed, you have a thing which if it is, it isn't.

MAN: A paradox circuit.

SPENCER-BROWN: A paradox circuit, yes. In putting it this way, this is the mathematics of it. I can put it in numerical mathematics, it's the same paradox. Make something self-referential, it either remembers or it oscillates. It's either what it was before or it's what it wasn't before, which is the difference between memory and oscillation.

WATTS: In introducing the word *before*, haven't you introduced time? You have a sequence.

SPENCER-BROWN: I have to apologize, because you realize that in order to make myself understood in a temporal and even a physical existence, as by convention is what we are in, remember I have to use words about the construction of the physical existence in order to talk about forms of existence that do not have these qualities. And if that were easy—this is one of the obstacles put in our way. Basically, to do what I am attempting to do is impossible. It is literally impossible, because one is trying to describe in an existence which has them—one is trying to describe in an existence which has certain qualities an existence which has no such quality. And in talking about the system, the qualities in the description do not belong to what we are describing. So when I say things like, "To oscillate, it is not what it was before; to remember, it is what it was before, I am describing in our terms, something that it don't have. But, by looking at them, you can see.

Mystic "Nonsense"

SPENCER-BROWN: This is why in all mystical literature, people say, "Well, it is absolute nonsense." It has to be absolute nonsense because it is attempting to do this. But it is perfectly recognizable to those who have been there. To those who have not, it's utter nonsense. It will always be utter nonsense to those who have not been to where the speaker is describing from.

The theory of communication is absolute nonsense. There is no reason whatsoever why you should understand what I am saying, or why I should understand what you are saying, if I don't recognize from the blah, blah, noises coming out of your mouth, that mean nothing whatever, where you have been. You make the same noises that I make when I have been there, that is all it is.

For example, Rolt, in his brilliant introduction to the *Divine Names* by Dionysius the Areopagite, begins describing the form at first, and then he actually describes what happens when you get the temporal existence. It is all the same thing, but he is describing it in terms of religious talk, theorems become angels, et cetera. When he comes to the place, which he says most beautifully, having described all the heavenly states and all the people therein, et cetera, and he says, "All this went on in absolute harmony until the time came for time to begin"[4] This is quite senseless. But it is perfectly understandable to someone who has seen what happens, who has been there. One cannot describe it except like this. It is perfectly understandable. He had described the form and then he had done that, and this is the time for time to begin.

Mathematics and Its Interpretations: Nots and Crosses

SPENCER-BROWN: There is just one question that I have been asked to answer, and I think it is something that you, Gregory, asked, wasn't it? to do with *not*. Was the cross—the operator—was it *not*. No it ain't.

---

4. What Rolt actually writes is somewhat subtler: "…until the appointed moment arrives for Time and the temporal world to begin" (Rolt, 2015, p.9. In Dionysius the Areopagite. [2015]. *On the Divine Names and the Mystical Theology.* Whitefish, MT: Aeterna Press.).

If I can, I'll try to elucidate that. I am reminded of one of the last times I went to see Russell and he told me he had a dream in which at last he met "Not." He was very worried about this dream. He had a dream, and he met "Not," and he couldn't describe it. But by the time we are using logic, we have in logic *not*.

We say: "*a* implies *b*." I am assuming that we know the old logic functions. You can describe this, ¬*a*, as *not a*. Now that *[writes: ¬]* is not—that is a shorthand for *not* in logic. *Not* in logic means pretty well what it means when we are talking, because after all, logic is only mildly distinguished from grammar. Just as we learn after reading Shakespeare's sonnets that after all they are full of grammar. Some people seem to think that all we have to do is learn grammar to be able to write like that—not so. So, they're full of grammar—they're also full of logic.

Grammar is the analysis of the constructions used in speech, and logic is the analysis and the formulation of the structures and rules used in argument. Now in arguments, there are the variables, "if it hails, it freezes," and the forms; We can say, in that case, it means the same thing as "either it doesn't hail, or it freezes," and find this is actually what *implies* means. We can break down *implies* into *not* and *or*.

Now when we are interpreting whenever are using the mathematics, we write *a* for "it hails," and *b* for "it freezes." If it hails, then it freezes; either it doesn't hail, or it freezes. And in the primary algebra we can write, "*a* cross *b*." $\overline{a]}\ b$. *[Writes statement on board.]* The primary algebra does not mean that. We have given it that meaning for the purpose of operation, Just as we may take a whole system of wires, electric motors, et cetera, and we can put it into a mathematical formula, or we can take some cars and weights, et cetera, and put them in one of Newton's formulae for finding acceleration. But the formula is not about cars, and so on and so forth; nor is this formula about statements in logic.

Just as here we have used *a* to represent the truth value of the sentence, "It hails," and *b* to represent the truth value of the statement "It freezes," we are in fact applying, because we recognize the structure is similar, the states of the first distinction to the truth values of these statements. We recognize the form of the thing. And in fact, *not*, in this case, although it is represented by the cross, the cross itself is not the same as *not*. Because if it were—well, we can see obviously that it isn't, because, in this form we have represented true by a cross and false by a space...if you represent true by a space and false by a cross, then wherewith our *not*? We have swapped over and identified the marked state with untrue this time, and the unmarked state with true. And here we have identified it with untrue. Change over the identification, which we may do, and now here is the statement. And if this were *not*, this would now have two *nots*—but it is not *not*. We have only made it representative of not for the purpose of interpretation, just as we can give a color a number and use that in altering an equation. But the number and the color are not the same thing. This is not *not* except when we want to make it so. But it has a wider meaning than *not* in the book.

WATTS: Well, it means that it is distinct from.

SPENCER-BROWN: No, no—it means cross.

Marked State/Unmarked State

SPENCER-BROWN: If you go back to the beginning of the book, you see—you remember this is not what really happens, because nothing happens. We represent what doesn't actually happen but might happen if it could. We represent it in the following way: We may draw a closed curve to represent a distinction, say the first distinction. Now we have a form. And we will mark one state, so, in fact. The mark is, in fact, shorthand for something like that, because it is only a bracket we marked it with. If we don't mark it with a bracket, we find that we have to mark it with a bracket, as I show in the notes.[5]

WATTS: Well, you have got it in the frame of the blackboard.

SPENCER-BROWN: Never mind about that. Now, let there be a form distinct from the form. Let the mark of the form be copied out of the form into such another form. Let any such copy of the mark be taken as a token of the marked state. Let the name as the token indicate the state. I missed out a sentence. Let the token be taken as the name, and let the name indicate the state—right.[6] Now, here, this indicates the state. We now derive our first equation from Axiom One—if you call a name twice or more, it simply means the state designated by the name by which you call it. So we have the first equation: $\overline{}\,\overline{} = \overline{}$. *[Writes equation on board.]*

Then, let a state not marked with a mark be called the unmarked state, and let any space in which there is no token of the mark designate the unmarked state. in other words; we did away with the second name. This is essential. It's the fear of doing away with the second name that has left logic so complicated. If you don't do away with the second name, you can't make the magic reduction.

BATESON: Are you saying that the name of the name is the same as the name?

SPENCER-BROWN: No, no, no. I said here, if you call a name twice, it is the same as calling it once. Your name is Gregory. If I call you twice, it is still calling you.

What people have done is that they have given a name always to both states. There is no need to do that; you have got quite enough to recognize where you are, because you do a search, and if you find the mark, you know you are in the marked state. If you do a search and you don't find it, you know you are in the unmarked state. So that, mathematically, is all that is necessary. So you don't do the second thing. There has been always fear, you see, to have a state unmarked.

---

5. To Chapter 2, *Laws of Form*
6. *Laws of Form*, p. 4. The actual sequence is: "Let there be a form distinct from the form. Let the mark of distinction be copied out of the form into such another form. Call any such copy of the mark a token of the mark. Let any token of the mark be called as a name of the marked state. Let the name indicate the state."

MAN: How did the printer feel about this. It must have driven him crazy.

SPENCER-BROWN: Oh, he didn't like it—he kept putting things in. The printer and the publisher went absolutely haywire because of equations like this: *[Writes on board:* [⊐ $_=$]*.]*

MAN: The American military documents, because of the number of pages that have to be printed, frequently have a blank page, And to be sure that nobody gets confused about it, there is always a statement on that page that says, "This page is deliberately left blank," which, of course, it is not.

SPENCER-BROWN: You see, why it has taken so long for the laws of form to be written is that one has to break every law, every rule, that we are taught in our upbringing. And why it is so difficult to break them is that there is no overt rule that you may not do this—why it is so powerful is that the rule is covert.

There is no rule that is overt anywhere in mathematics which says this may not happen, it may not be done. And it is because I found no such rule that I gathered that it could be done, and that it must be done. If you don't do it, you are not doing the mathematics properly, and that is why it is all such a mess. This is only a social rule that you may not do it. And there is no mathematical rule that you may not do it; and in fact, you have to do it. Otherwise the mathematics is a mess and you can't get the answers because you are blocked.

Now to go on to what I was going to say, which is: next we want to use the mark, which could be a circle. We want to use it. We haven't, in fact, discovered its shape. In the second equation, we discover, really, what the shape is. And we'll see it is inevitable. Having marked one side *[draws a letter* m *on the board]*, if there is no mark, then we know we are on the other side.

CLIFF BARNEY: Those *m*'s are outside the circle or inside the circle?

SPENCER-BROWN: Well, this one is outside.

BARNEY: I wasn't clear on which is the inside.

SPENCER-BROWN: In fact, it depends on where you are. This is already beyond what we have said mathematically, because, in fact, this is only an illustration. Just as, when you play Beethoven's music it is only an illustration of what Beethoven wrote. All mathematics in books is only an illustration of what cannot be said. This illustration is misleading because there is no outside or inside when you have drawn the first distinction. You have just drawn a distinction—we can illustrate it with a circle because it happens to be convenient. Then we mark one side, and we know, in this case that it is the outside. But remember that in the mathematics there is no outside. There are just two sides. We have marked one of them, and if we find the mark, we know we're in that state. We call it the marked state because it is convenient to call it by something, which, having marked it, we'll say it is a marked state. Simply for something to call it. And having not marked the other side, we call it the unmarked state. That is all that is needed. We now have every concept we need.

First of all, we have taken the mark as a name. And if you call a name twice, you are simply indicating the same state twice, and indicating the same state twice is the same as indicating the same state once. Now, instead of just calling this *m*, let us give it certain properties. Let it be an instruction to cross the first distinction.

Now here is our illustration of the first distinction.[7] Now this is why we've drawn this line on our blackboard, because here is an illustration of the first distinction. Here is a record of instructions referring to the first distinction—right. Now let *m*, the mark, be taken as an instruction to cross the boundary of the first distinction. So, if one is here, *m* says go there. If one is here, *m* says go there. OK? *M* is now not a name, so we can ring that for the moment, don't confuse yourself with that, *m* is now an instruction. And all *m* means is cross. So whenever you hear or see it, you've got to step over the boundary. That is all it means. Now, we will produce more conventions.

We will say that we have got a number of crosses considered together, and these we will call *expressions*. Now suppose you have this. We'll say—right—we'll represent *m* like that. And we'll say *m* means cross and we'll make a convention so that whatever is represented in here, you'll have crossed to get what is represented out there. So if there is nothing represented here, absence of the mark indicates the unmarked state. You cross when you are in the unmarked state and you find you are in the marked state. So out here by representation will be a value attributed to this mark, will be the marked state, and that is the value we attribute to that expression.

Now let us put the marked state in here, and we can do that simply by putting another cross in here. Now the convention is that wherever you see nothing you are in the unmarked state. Wherever you see this, you must cross. So, here we are. We hear nothing, we see nothing, we are in the unmarked state. Our instructions now say "cross," so we cross, and then our second instruction says "cross," so we cross. So here we are, we started here and we have crossed, and we have crossed here, and so we can derive our second equation: *[Writes on board:* ⊐⊐ = . *]*

So that all this says in mathematics is "cross." It does not say "not." It says "cross."

End of Session One

---

7. For illumination of what follows, see pp. 82–84 in the notes to Chapter 2, *Laws of Form*.

AUM Conference: Session Two
Monday Afternoon, March 19, 1973

*transcript created and edited by Cliff Barney
from recordings by Kurt von Meier*

SPENCER-BROWN: I am aware of a number of different pulls as to which way we could go from here, and in a universe where there is a degree of exclusion, one way excluding the other in a finite amount of time, I'd like to get some consensus as to which way we might go from here. If we could just ask questions of people as to what we could profitably talk about next.

Algebra and Arithmetic

VON FOERSTER: I think it would be lovely if you would make again for us the very important distinction between algebra and arithmetic. Because the concept of arithmetic is usually—although every child knows about it, and it is plain everyone knows about arithmetic—and here arithmetic comes up in a more, much more, fundamental point. And I think, if this is made clear, I think a major gain will be made for everybody.

SPENCER-BROWN: I'll do what I can to make the distinction clear. I was going to say, "make the distinction plain," which means to put it on a plane. I suppose that most people know that the meaning of the word *plain*, if you look at its root, is just another word for plane, plane like blackboard.

To make plain is to put it on a plane. So that's what I will do. I will try to put this distinction between algebra and arithmetic on a plane. The reason it should go on a plane is that in a three-space it is difficult to disentangle the connections. So we project it onto a plane. On a plane, we can take a plan, which is the same word.[1] We can see the relationships of the points on inner space, which is not too difficult to comprehend.

Now to make a distinction between algebra and arithmetic, I should go to the common distinction which is made in the schoolbooks, where you have—there are two subjects in kindergarten, perhaps a little beyond kindergarten—a subject called arithmetic, which is taught to you first, and then we have algebra, which is what the big boys and girls get onto and look rather superior about. First of all, let me explain—this word keeps coming up, "out on a plain"—that even arithmetic is not what is taught at school. Mathematics certainly isn't. What the child is first taught is

---

1. Practically the same. They are actually derived from two different Latin words, one (plane) meaning flat surface and the other (plan) meaning sole of the foot; and before that, respectively, from two different Indo-European words, one meaning flat and the other meaning to spread.

the elements of computation—the computation of number, not of Boolean values. He is taught the elements of computation, which is wrongly called arithmetic. Whereas arithmetic is the…

Let's be clear, for the moment. I'll go back and start again. We should approach this slowly and deviously. I don't want to give the game away before we have got there. In arithmetic, so-called, which the child is—it is true that it doesn't begin with arithmetic, because the child is given an object, two object, three objects, four objects, and he is—I don't think that he is taught that there is something called a number, but he is then taught to write "one, two, three, four," et cetera, and he is not given that there is somewhere between that—that is one object, that is two objects and that and that, that, that. There is somewhere between these a non-physical existing thing called a number. As I point out in *Only Two*, a number is something that is not of this world.[2] That doesn't mean to say that it does not exist—it surely does. In fact, there are many extant groups of numbers.

VON MEIER: Exstasy? Exstasis.

SPENCER-BROWN: Well, yes, that's *outstanding*. Existence, *ex*, out, *stare*, to stand, outstanding. What outstands, exists. And numbers do not outstand, they do not exist in physical space, They exist in some much more primitive order of existence. But they, nevertheless, do exist. But not in the physical universe.

This, by the way, is the first way to confound the material scientist who thinks that physical existence is all there is. You ask him—"Well, you know that there are numbers?" He will perhaps have to say that there aren't any numbers; in that case, you can't beat him. If he admits that there is such a thing as a number, then you say, "Well, find it, where is it, show it to me," and he can't find it. It does not exist here.

Now, a child may come to learn very much later, here, there are objects arranged in groups, here are figures. Number is to be found in another space. Not in this space. However, these are the symbols, tokens of number, which can be in any form—Roman, et cetera. Playing around, saying "two plus three equals five," an elementary computation with numbers, is discovering relationships with numbers, and how they are constructed and what they do together. Sounds a bit rude, but that's what we do.

When the child gets a bit older, he is taught what is called algebra. The first teaching of algebra that was given to me, it may be the same here, was that we were given things like "a plus b equals c; find c when a equals five and b equals twelve." And we all scratched our heads and learned to do this sort of thing. Eventually we came to formulae that were algebraic, and were finally told things that were universally true. We were taught that an algebraic relationship is true irrespective of what numbers a and b stand for. In other words, as we learned algebra, we learned it as an extension of arithmetic.

---

2. See Keys, 1972, *Only Two Can Play This Game*, Note 4, pp. 134–135.

As we got a little older still and went to the university, we learned different names; and we were taught that, whereas, these were *constants* and these were called *variables*, you could learn the science of algebra without ever knowing what those words stood for at all, treating algebra as a possible system, and having derived, actually, your rules of what to do, in the case of an ordinary algebra of numbers, from experimenting with the arithmetic. Eventually you see what the rules are, and you operate and find things out without referring back to the constants.

I have given the game away now. This is the difference between an algebra and an arithmetic. The algebra is about the variables, or is the science of the relationships of variables. It is a science of the relationships of the variables when you don't know or don't care what constants they might stand for. Nevertheless, the constants aren't irrelevant, because whatever arithmetic this is an algebra of, if you were to substitute constants for these variables, a, b, et cetera, then these formulae still will hold.

A lot of people have said, you see, "How can you have an arithmetic without numbers?" as the primary arithmetic in *Laws of Form* is without numbers. We will go back in a moment to that. But just at the moment we will emphasize, or return to, for memory purposes, the fact that the definition—the difference between algebra and arithmetic is that arithmetic is about constants, the algebra is about variables. The arithmetic is a science of the relations of constants.

A common arithmetic for university purposes, which for a less vulgar name is called the theory of numbers, is the same thing. The theory of numbers is arithmetic, it's common arithmetic. The theory of numbers, the most beautiful science of all in mathematics—I happen to like it myself, so I praise it—or one of the most beautiful, is the science of the individuality of numbers. A number theorist knows each number in its individuality. He knows about the relationships it forms, and so on, as an individual, as a constant. An algebraist is not interested in the individuality of numbers, he is interested in the generality of numbers. He is more interested in the sociology of numbers. That applies, whatever individual numbers come there; he has produced a rule where these people go there and there and there, and so on, and he's not interested in individuals at all.

A very interesting point here is the illustration of Gödel's theorem in the difference between, in number theory, an algebraic factorization of a number and an accidental factorization. As you know, we know from Gödel's theorem that in the common arithmetic, that's the arithmetic of the integers, the algebraic representations, the rules of the algebraic manipulation of numbers, do not give you the whole story. It doesn't give you the complete story of what goes on in arithmetic. And so we have this factorial relationship—any number that is in that form, we know will factorize into that form. But there are what are called in number theory *accidental factoralizations*, which happen over and above and irrespective of any algebraic factoralizations that you can find. And this is a very beautiful illustration of Gödel's theorem, Nobody has ever used it. I think this is because, in general, mathematicians don't understand Gödel's Theorem or even know what it says. I have lectured to an audience of maybe fifty university mathematics teachers. "Can anybody tell me

Gödel's theorem?" Not one. Not one knows what it is. It is one of the extraordinary breaks which mathematics took about the turn of the century. Where logic broke off from mathematics, and the two, you know, despised one another; like in gliding and power flying, they weren't speaking the same count. Hence we have this tremendous break, this schizophrenia, in mathematics, where common illustrations of one thing in another field just aren't seen as such. Accidental factoralization is a most beautiful illustration of Gödel's theorem, if a somewhat technical one, in number theory.

Now having seen, therefore, the difference between algebra and arithmetic—simply that arithmetic is concerned with constants and algebra is concerned with variables—we have, well, as Whitehead points out in the *Treatise on Universal Algebra, Vol. 1*, he points out that Boolean algebra is the only form of non-numerical algebra known. Shortly after that, there was a book written by Dickson, who is also a number theorist of some considerable fame, who wrote a very wonderful book called the *History of the Theory of Numbers*, now published by Dover; and anybody who is interested I think should get it because it contains all that would be of interest, except a very few later things. And he starts right at the beginning with amicable numbers,[3] and shows that the early mathematicians, if they wanted to be friends with somebody, would find a pair of amicable numbers, and they would then swap numbers and they would eat the number of their friend, to keep the friendship. All this is in the *History of the Theory of Numbers*, by Dickson, which is a wonderful book. He also wrote a book called *Algebras and their Arithmetics,* which—I don't think he actually said it, but it was obvious that every algebra has an arithmetic.

At the same time, mathematical popularizers such as W. W. Sawyer were writing popular expositions of various forms of mathematics, including Boolean algebra. And Sawyer heads his chapter on Boolean algebra, "The Algebra Without an Arithmetic." This can't be. This can't be. If it is an algebra, it must have an arithmetic. And if any mathematician could write this—I am not blaming Sawyer; Sawyer was only standardizing what is common mathematics taught in universities today. He is standardizing the common confusion and block. The fact that mathematics teachers in universities today do not understand the difference between an algebra and an arithmetic, which is simple.

How to Find Laws of Form

SPENCER-BROWN: So, to find the arithmetic of the algebra of logic, as it is called, is to find the constant of which the algebra is an exposition of the variables—no more, no less. Not, just to find the constants, because that would be, in terms of arithmetic of numbers, only to find the number. But to find out how they combine, and how they relate—and that is the arithmetic. So in finding—I think for the first time, I don't think it was found before, I haven't found it—the arithmetic to the algebra of logic—or

---

3. Those whose divisors—other than themselves—add up to the same number.

better, since logic is not necessary to the algebra, in finding the arithmetic to Boolean algebra—all I did was to seek and find a) the constants, and b) how they perform.

And the first four chapters of *Laws of Form* are just about this arithmetic, And the nine theorems, with which the two connective theorems later form what would be called in any other algebra *postulates*, are called here *theorems*, because they are proven. They are not postulates—you do not have to postulate. These are the basis upon which we can build the algebra. The initial equations…are the rules of the arithmetic—or rather, they are all the equations necessary for the arithmetic.

VON MEIER: What geometries would follow from this?

SPENCER-BROWN: None whatever.

VON MEIER: Would there be any relationship?

SPENCER-BROWN: Well, geometries are sciences of spaces of this kind, of a three-space or something like that, where you already have measurement. In the initial space which we are concerned with, where you have just drawn the distinction, there is no size and therefore no measurement, and no geometry can follow from it. Or, if you like, this is the geometry of it. In other words, we are in a place where geometry and arithmetic condense. Later on we can see, of course, in the Euclidean geometry, for example, that we can express it algebraically and, therefore, arithmetically, without figures.

VON MEIER: This is a plan of geometric icons, then.

SPENCER-BROWN: Well, the point is that you don't have any geometry as distinct from the arithmetic here, because if you go into the definition of mathematics in a textbook, you will see that it is the science of spatial relationships—it's about space. The simplest science, the simplest form of space, is of distinction.

Proof and Demonstration

KARL PRIBRAM: Is this related to the difference between demonstration and proof?

SPENCER-BROWN: Ah, the difference between demonstration and proof is that a demonstration is always done by the rules. A computer can do a demonstration. We can demonstrate this ⌐⌐⌐, condense, we get ⌐⌐ and then condense ⌐ to—that goes out and we get that and that goes out, you see and it goes to nothing. And here ⌐⌐⌐ we just cancel twice and it is nothing. So you demonstrate that—there, that's demonstrated—that's OK, you see. And we can demonstrate algebraic forms. This is a demonstration of arithmetic and a demonstration of algebra as well.

Now proof is quite different. Proof can never be demonstrated. I will give an example of proof—one which is familiar; I am always giving it as an example and those who have been given this example will forgive me if I give it again now, because it is a very beautiful theorem, a very beautiful proof by Euclid to show the difference between a proof and a demonstration. I go to the illustration rather than to the Boolean form, I go to the common school form, the arithmetic of numbers, which is so familiar to us all, and therefore, I think, better for illustration.

We can give a demonstration, a computer can demonstrate, we just follow the rules within the calculus. Where we have to prove something, we always have to find—we cannot find it with the rules within the calculus. In other words, no computer will compute a proof.

Prime Numbers

And we take, for example—or for the purpose of illustration—Euclid's proof of his beautiful theorem, the question asked, "Is the number of primes infinite?" As we see, the prime numbers, and it's obvious when you think of it, as they go on, they get sparser. It's very obvious that they will, if you consider it, because every time we have a new one, we have a new divisor which is likely to hit one of the numbers we're looking for to see if it's prime. If it hits it, if it divides into it, then it won't be prime. So, the bigger the number, the less likely it is to be prime. A strange sort of statement. the science of certainty, taken in probability terms. Because the more primes there are that could divide into it. So for fairly obvious reasons, as we continue in the number series, the primes get, in general, further and further apart. there are fewer and fewer of them. And what Euclid asked was, do they get so thinly scattered that in the end they stop altogether? Or does this never happen?

This is an example, now, of a mathematical theorem. To make it into a theorem, you actually give the answer, you actually state the proposition, "The number of primes is endless." You may not be certain whether it's true or not; you may still be asking the question: "Do they come to an end or do they go on?"

Well, to illustrate the difference between mathematical art, because it now needs an art to do the theorem, where it only used a technique, a mechanical application, to demonstrate something, and we don't need to do it ourselves, as computers can do it so much better, we will now do something that a computer can never do. Because what we are going to do is find the answer to this question—do the primes go on forever or not? We are going to find this answer quite definitely, and we are not going to find it by computation, because it cannot be found by computation; but it can be found like this. This is the way Euclid found it. He said, supposing they come to a stop—all right, if they come to a stop, then we know they are going to go on for a long time until we come to big primes, but, if they do come to a stop, there will be some largest prime, call it Big N. That's it. That is the last prime, the biggest of the lot. If they come to a stop, there must be such a prime. Now, if there is such a prime, and there it is up there, let us construct a number which looks like this: all primes, every single one of them, up to and including Big N. Right. We have made this number by multiplying all the primes together Now, Big N being the largest, this is a number which is made of all the primes there are, there isn't another prime. Because we have assumed that this is the largest.

On the hypothesis that this is the largest, this number is now all the primes multiplied together, and we'll do this multiplication and get the answers and we'll call the answer Big M. We'll take this number Big M, and we will add one. Now we will examine the properties of Big M Plus One. You see this is why arithmetic is so lovely: it is about individuals. Here is our number Big M, as an individual; here is Big M Plus One. It is a hypothetical number, actually, it is a nonexistent number—this is why we can't speak of numbers not existing, because some of them do and some of them don't. Big M Plus One, let's examine its properties. Well, it is obviously not divisible by any other prime, including Big N, because we know they all divide Big M; therefore every single prime leaves a remainder of one when we attempt to divide it into Big M Plus One. So Big M Plus One, therefore, must either be prime, because it is not divisible by any existing prime; or if it ain't prime, then it must be divisible by a prime which is larger than Big N. Therefore, by assuming that there is a biggest prime, call it Big N, we have ineluctably shown that this assumption leads, absolutely without any doubt, to the construction of a larger prime, which is either Big M Plus One or another dividing Big M Plus One. And that is how Euclid did it, and that is—there are many other proofs, of course, but it still is one of the simplest and most beautiful, and the answer is absolutely certain that there is no largest prime, that they do go on forever. This cannot be done by a computer. Currently there is no computer that has done that.

BATESON: There must be intervening primes, you might say accidental primes, just like that accidental factorization business.

SPENCER-BROWN: Between when? Where?

BATESON: Between primes that are made by multiplying sequences of primes and adding one.

SPENCER-BROWN: It is not necessary to make primes, you see. This is not necessarily prime, you see.

BATESON: It is not necessarily prime?

SPENCER-BROWN: No, it isn't.

MAN: Multiply three by five, add one, that's 16. Non-prime.

SPENCER-BROWN: We have to add two. You get 33 and it's non-prime.

BATESON: Non-prime. Why in heaven's—

SPENCER-BROWN: It doesn't have to be prime, you see.

MAN: Thirty-one Thirty-one! Not 33. Thirty-one is prime.

SPENCER-BROWN: Right, we get 31. But if you go out far enough, you will find that you get one that isn't prime. But they will be divisible by a prime bigger than the largest prime you have used. Let's see if we can find one. The—ya, here, wait a minute, 211 is prime, isn't it? I'm just thinking of the prime factorial plus one; at seven, it's two, one, one. That's prime factorial plus one. 211 is prime, as far as I know.[4] We want a table of primes here. Not divisible by 13, is 4...we multiply the next one,[5] 11, 2, 1, 1, 2, 1, 2, 3, 2, 1. Sorry, 11, 2, 1, 0, 2, 1, 0, 2, 1, 3, 1, 0. And so it comes out 2311. Is that prime? Probably not. I am very bad at figures.[6] Divisible by 13...4...not divisible by 13....

Anyway, I do assure you that if you go on long enough, getting the final factorial, adding one, you will find one that is not prime; but that doesn't matter, because it will be divisible by a prime that is bigger than the biggest prime you have used to produce it.[7] If it were always prime, you would have immediately a means—you would have a formula for producing primes, and this we haven't got. There is no formula for producing primes except going about it the hard way and seeing as they don't divide by anything.

Theorems and Consequences

SPENCER-BROWN: Now, this is totally confused, the idea of the difference between demonstration and proof in mathematics. In fact, Russell, you see, in suggesting it, completely confused them, and people have done so ever since. What he called theorems are in fact consequences, they are algebraic consequences, which can be, in fact, demonstrated. And indeed, he says, "These theorems"—he calls them theorems, they are consequences—"can be proved." And then he does the demonstration and then he calls it *Dem*. *Dem* is short for *demonstration*. The two words are used interchangeably, and wrongly. There is a difference, and what can be demonstrated is done within the system and can be done by computer. And what cannot be

---

4. It is.
5. Meaning the next prime, 11.
6. 2311 is indeed prime
7. *2311* is only one of the names assumed by Big M Plus One. This number is constructed by adding 1 to 2310, which is "prime factorial" for 11—that is, the product of all the primes up to and including 11: 1 x 2 x 3 x 5 x 7 x 11. Under this alias, Big M Plus One is prime, not divisible by any numbers except itself and 1. In its next manifestation, however, as 30,031 (prime factorial plus 1 for 13), our space messenger is composite. Thus it confirms Brown's assurance, that eventually, "if you go on long enough, getting the final factorial, adding one, you will find one that is not prime; but that doesn't matter, because it will be divisible by a prime [in this case, two, 59 and 509] that is bigger than the biggest prime [here 13] you have used to produce it."

demonstrated, but may be proved, cannot be done by computer. It must have a person to do it. No computer can prove it, because it is not proved by computation. The steps of this proof, Euclid's proof, were not computational steps. No one could do it on a computer, because we were not doing computation. We were divining the answer, we were divining what had to be done by making certain deductions and seeing what they led to. This was an artistic process, not a mechanical one.

The computer cannot do it because it is not computation. Computation is counting in either direction, no more, no less. There is nothing more to computation than that, nothing more.

MAN: I am trying to determine what it is that a human can do that a computer can't do.

SPENCER-BROWN: Let's go through the steps again. Where is the computing? We compute nowhere. There is no computation in this proof. Not a single computation can be made, not one. The whole process is a proof. In the whole process of a proof, there is not one single computation, nothing that a computer could do.

MAN: Well, there are two fake computations.

SPENCER-BROWN: There are no computations. They were fakes because there were no such numbers. We were imagining doing a computation of a particular kind—we weren't actually doing it, because there were no numbers to put in the places. In fact, there only could have been a computation if our number Big N, being prime to the largest, happened to exist. Yes. If it happened to exist, and we knew what it was, then we could do this whole thing on a computer. But it doesn't happen to exist. But in order to find it doesn't happen to exist, we go through the imaginary steps of computing in this particular way, and then we find that if we did that we would find another number which is prime or contains a larger prime.

MAN: John Lilly, what does the biocomputer offer as the possibility of doing this, if the computer doesn't?

LILLY: Well, the biocomputer does the whole thing.

MAN: Say it again.

LILLY: The biocomputer invented the whole thing.

SPENCER-BROWN: I know, as an engineer, the computer boys have vastly oversold their products by saying that they can do anything that the human mind can do, and this is not so. They cannot do the most elementary things that the human mind can do. And I blame...I blame Russell/Whitehead for totally mixing up proof and demonstration. If you go through *Principia*, there is not a single theorem, not one theorem. I think I am right about that, Dr. Von Foerster, because what they call theorems are consequences.

Also, they had a precedent in that Euclid himself already rightly called this a theorem, calls it algebraic. His geometric consequences, he called theorems, they are not. So the confusion developed right at the beginning with Euclid, who called his geometric consequences, they can be computed, he called them theorems. Wrongly. So Euclid was the first offender. And from him, it just shows how we copied. We have copied his error through hundreds of years.

VON FOERSTER: I think the Q. E. D. thing makes it appear as though it were a demonstration—*quod erat demonstrandum*. It should not have been called demonstrandum.

SPENCER-BROWN: I may be wrong, you see. My Latin—I have little of it—perhaps he was OK. He said *quod erat demonstrandum*, "this has been demonstrated." It is OK after a demonstration, it is misleading after a proof. And maybe he did not make this error, but we have. We have called them theorems when we should call them consequences. And this has been responsible for—a vast system of error has grown up there. Because a computer has been found to be able to demonstrate consequences—because all you need is the calculating facility to do this. And consequently the demonstration of consequences, in other words, calculations, has been confused with the proof of theorems, which is another matter altogether. Because of this confusion, it has been thought that a computer therefore can do practically all that a man's mind can do. But it can't, because only the most minor function of a man's mind, done very badly, is to compute. And we have, in fact this tremendous emphasis, because of the confusion in mathematics—the difference between computation and actual mathematical thinking—which has led us to believe that computers have minds, can do what we can do.

For example, they put Russell's consequences on the Titan computer at Cambridge. It managed, with great hesitation and very slowly, to demonstrate a few of them, but the more complicated of them it couldn't demonstrate. However, it could have done it, eventually. It was very slow and expensive. Even here, what a computer can do, a man can do better if he gives himself to the problem, because he has the capacity of seeing in a way the computer never can.

MAN: Can you say what that capacity is? That makes us different?

SPENCER-BROWN: Can I say what it is? No, I can only represent it. I can only be it. Just as you are. How can one say what it is except to give examples: Computing is 1, 2, 3, 4, 5, space, space, space, space, space, 6, 7, 8. That's five plus three is eight. that's how a computer does it.

[Tape becomes unintelligible at this point.]
SPENCER-BROWN: I have been asked earlier if I would go through with you the main mathematics of the book, and I think this is not possible because there is not enough time, and it is so varied an audience.

I did do this in London, but it was a series of 20 lectures. I gave it seasonally every year, and even then it wasn't enough to do anything but in outline. Only in the last two or three lectures was it possible, having got most of the audience to an understanding of what it was about, was it possible to show how it was related—how it could possibly relate to the disciplines of everyday life.

Unless there is a very strong expressed desire here, I don't feel that it would be terribly desirable, for the majority of people here, if I did actually go through technically and get as far as one could—because it wouldn't be far enough to draw on the conclusions that are possible after more detailed study. I don't think it could be done.

It's just not possible to do everything all at once. You can't make the rice grow by pulling on the stalks. If there is anything further that you would like to discuss now, we'll see what we can do. Or would you like me to suggest one?

First Distinction, Observer, and Mark

LILLY: At the end of Chapter 12, you make a sort of covert statement. You do not develop it, and I'd like you to develop it a little further. You mention that it turns out that the mathematician is one of the spaces.

SPENCER-BROWN: The mathematician?

LILLY: Yes, is one of the spaces.

SPENCER-BROWN: The part of the observer? "We now see that the first distinction, the observer, and the mark, are not only interchangeable, but, in the form, identical."[8] I don't see how you didn't get it already. The convention that we learn to grow up by is—the game that we are taught to play is that there is a person called *me* in a body called *my body*, who trots about and makes noises and looks out through eyes upon an alien, objective thing we call *the world*, or, if we want to be a bit grander, called *the universe*, which the thing called *me* in *my body* can go out and explore and make notes about and find this, that, and the other thing, find a tortoise, and make notes about a tortoise, and drawings, et cetera. The convention is that this tortoise is somehow not me, but is some object independent of me, which I in my body have found.

We also have a further convention—well, depending on what sort of people we are, if we are behaviorists, we may not think this—most of us think that the tortoise also sees life in much the same way, that it is a being that has *my shell, my feet, my tail,*

---

8. *Laws of Form*, p. 76. See also Blake couplet quoted on p. 126, *Only Two Can Play This Game*: "If you have made a circle to go into/Go into it yourself and see how you would do."

my head, my eyes, out of which I look through the hole in the front of my shell and I see objects, big things walking around on two feet, et cetera, which are different from me. And we think that the tortoise thinks that.

Now, supposing that this is only a hypothesis. Supposing that—if there were a distinction—if there were that—only supposing that, if it could be, what would happen—well, if one imagined—supposing one imagined, well, this is me and that ain't me. Surprise, surprise, what ain't me is exactly the same shape as what is me. Surprise, surprise.

Come to this another way. Take it philosophically. Take it philosophically and scientifically. Scientifically, on the basis of, "There is an objective existence which we can see with our eyes and feel with our fingers and hear with our ears, and taste with our tongues, smell with our nose," et cetera, and then we take it to ants. Now ants can see ultraviolet light, which we can't see, and therefore the sky looks quite different to them. Take it to extremes. If there are beings with senses, none of which compares with ours, how could they possibly see a world which compares with ours? In other words, even if one considered it scientifically, the universe as seen appears according to the form of the senses to which it appears. Change the senses, the appearance of the universe changes. Ask a philosophical question and you get a philosophical answer. What therefore is the objective universe that is independent of these senses? There can be no such universe, because it varies according to how it is seen, the sensory apparatus. Take this a little further, and we see that we have made a distinction which don't exist. We have distinguished the universe from the sensory apparatus. But since the universe changes according to changes in the sensory apparatus, we have not distinguished the universe from the sensory apparatus. Therefore, the universe and the sensory apparatus are one. How, then, does it appear that it is so solid and objective looking?

Now, the answer to this profound question takes a lot of thought, but I will try to give it all to you in a very short time. Because it takes a whole series of remarkable tricks[9] before it can be made to appear like this. But since, if there ain't no such thing, then any trick within the *Laws of Form* is possible. This happens to be one of the possible tricks. If there is no such universe, if there is only appearance, then appearance can appear any way it can. You have only to imagine it, and it is so.

BATESON: Can you go into the proof?

SPENCER-BROWN: The proof, my dear sir, has nothing to do with the objective world, the proof is mathematical. Nothing in science can be proved.

BATESON: I see.

---

9. Re-markable, markable again (a whole series).

SPENCER-BROWN: It can only be seen. But where it is co-extensive with mathematics, in that, in fact, what is so in mathematics—The basis of what is so in mathematics is what can be seen. *Theorem* and *theatre* have the same root.[10] They are the same word. It is the spectacle[11] that we see, and the discipline of mathematics is to go to what is so simple and obvious that it can be seen by anyone. Without doubt, it can be seen. And from this—

BATESON: By turtles. Can it be seen by turtles?

The Turtle's Specialty

SPENCER-BROWN: I don't know whether turtles see it. If they do, they have a different discipline whereby they communicate it. We don't talk with turtles, and I can't answer that. I have never spoken to a turtle. But I am sure that turtles can see. Well, they can certainly contemplate reality. I don't know whether they need to see mathematical theorems. I don't know whether they play that game.

VON MEIER: Yes, they carry their numbers on their back—13 variations in the shells in a certain pattern. It's the second avatar of Vishnu, so that when you see the turtle, you're seeing it from the point at which the mythologists named it God. They have named the serpent the first avatar of Vishnu. She's the cosmic turtle swimming in the sea. And things that run around, run around on the back of the turtle.

LILLY: This is called "maya-matics."

Solid State

PRIBRAM: Why so solid? Why is objective, so-called reality, so solid?

SPENCER-BROWN: Well, it has to be, after all. Oh, dear, what we need is a 20-year course to get to that point.

JEAN TAUPIN: Any reality is real, the moment you perceive it as real?

SPENCER-BROWN: Well, *reality* means royalty. The words have the same root.[12] Whatever is real is royalty. And what is royalty but what is universal—the form of the families of England.

---

10. Greek *theasthai*, to view.
11. Latin *spectaculum*, from *spectare*, to watch, from Indo-European *spek-*, to observe. Related is Latin *speculum*, a mirror.
12. In Spanish, perhaps, where *real* means royal in English. But English *real* is derived from Latin *res*, thing, and before that from Indo-European *re-* to bestow. *Royal* comes from Latin *regalis,* from *rex*, king, and before that from Indo-European *reg-* to move in a straight line, rule.

VON MEIER: The measure, the rex, the regulus.

SPENCER-BROWN: Yes, that is true.

LU ANN KING: You said two things. The motive precedes the distinction. And then, later on, you said that one has to determine their relevance, that in searching for a clue, you also have to determine its relevance.

SPENCER-BROWN: You have to make a relevant construction, yes.

Separating Figure and Ground

KING: Well, just personally, what process do you—

SPENCER-BROWN: How do I do it? Just contemplate. One also tries all sorts of ways to get familiar with the ground. Why It may take you two years to find a proof which could be exposed in five, fifty seconds, is that you get familiar with the ground. You try in many ways that won't work, and then you try and try and try and then you realize that trying—one day you stop that. And then, almost certainly, you will find—or seeing it, seeing anything to be so...

I had been working on the second-degree equations for five years at least. I was thoroughly familiar with how they worked, and so on—hadn't seen what they were, theoretically. I was wondering whether to put them in Chapter 11, or some other beautiful manifestation of the form, whereby you break up the distinction and it turns into a Fibonacci series. Well, I won't go into that now because it is another thing altogether. In *Laws of Form* there is only about one twentieth of the discoveries that were actually made during the research. There is enough for 20 books, mathematically, and I had to decide what I could put out, and what I could put in. But the actual research in London is 20 times of what is in the book. And I wasn't quite sure whether to put it in at this point—because the book had to be finished. I wasn't quite sure whether to put in, with this chapter, this beautiful breaking up of the truth where you get the rainbow, which turns into the Fibonacci series. You break up white light and you get the colors. You break up truth and you get the Fibonacci.

VON MEIER: The logarithmic growth spiral?[13]

SPENCER-BROWN: Yes. I decided in the end that it was more practical to put in the expressions which went into themselves, because we did have practical engineering uses for this. But I still didn't recognize the theory.

We had been using it, my brother and I, in engineering, but we still didn't recognize what it was. So I sat down to write Chapter 11 and without thinking I wrote

---

13. The Fibonacci series goes 0,1,1,2,3,5,8,13,21...etc., adding each final pair of numbers to produce the next one.

down the title. I wasn't sure what I was going to call it, but I wrote something without thinking. I looked at it and I found what I had written was "equations of the second degree." Now, I was not aware of writing this down. The moment I had written it, that was—Eureka—that is what it was. The moment that I spoke of it to my brother and then to other mathematicians, it began to focus. Yes, of course. And then it was only a matter of an hour or so to go through and see the analogy, which I did on the blackboard this morning. To see the paradoxes and everything, all the same, all existent in the ordinary common arithmetical equations of the second degree. And this is what we were doing in the thrown-out theory of types; the coming to the knowledge of what it is.

How actually does this happen? It happened something like that, after five years of scratching one's head but thinking, nevertheless, let's find out more about it. And then it comes, in a way, quite unexpectedly; in a way, really, for which one can take no personal credit.

Educating the Child; The Covenant of the Cradle

RAM DASS: Is the five years the method to get to the space—from which all titles are, or there was an implication in what you said that your familiarity with the ground was the prerequisite for your then stopping trying and then out comes this thing, which is like a subliminal, or a latent, or something inherent in the analytical process—nothing more?

SPENCER-BROWN: Well, I will distinguish the proceeding. It goes very much like the education of the child. The child is born knowing it all, and it immediately has this bashed out of it. It's very disturbing. So it learns the game then. It learns the game that is played all around it—and with variations, it is much the same game in any culture, whether it is the ghetto, or ten thousand years ago, or today in America, or today in England, or today in China, or wherever it might be. It's much the same thing, with variations, of course, in the particular cultural pattern. It has its original knowledge bashed out—it must be bashed. Those of us who have gone back and remembered our births, remembered what we knew, and remembered the covenant we then made with those standing around our cradle, the realization that we now have to forget everything and live a life—

RAM DASS: Excuse me, is the word *know* the proper word to use? Doesn't that imply a knower and an object that is known? Couldn't you say that the infant was being it all?

SPENCER-BROWN: If you like, yes. I am only using words—you see, the language is no good for talking this way. We have to use these imperfect terms, which are based on distinctions. And you are quite right, it is not knowing, it is only in its interpretation, knowing. It is like dreaming a funny dream. While the dream is going on, it isn't funny. But bring it out into the critical atmosphere of waking life—now it appears funny. The child is bringing out into this, and it remembers it as knowing it. That is the way it is taught the disciplines.

Can't Have One Without the Other

RAM DASS: But you never can get into knowing it. You can only get back into being it again. You can only know a segment or a—

SPENCER-BROWN: Well, it's dual, of course, because getting back from the—there is no enlightenment without un-enlightenment.

RAM DASS: There is no survival without un-enlightenment, actually.

SPENCER-BROWN: Well, I'll come to that.

VON MEIER: The planted seed is always regarded by primitive cultures as having died.

SPENCER-BROWN: Enlightenment is different for every form of culture, because every form of culture is a form of un-enlightenment. And the enlightenment matches it, as the form of enlightenment for our culture matches our culture. It matches the way in which we have been unenlightened. Enlightenment by itself, there is no such thing, just as there is no black without white. But to be enlightened, having been un-enlightened, is not the same as having been un-enlightened before. Because one wasn't really un-enlightened at all.

MAN: We need another word.

SPENCER-BROWN: Ya. Before, you are neither enlightened nor un-enlightened. Then you become un-enlightened, from which you have to be enlightened. That is not the same. You remember your original un-enlightenment.

PRIBRAM: Original lightenment.

SPENCER-BROWN: Well, that would be, in a way, but it might hurt.

VON MEIER: When one sees the light for the first time, from the interior of logical models, or from the cosmic tortoise in the sea, or from the inside of the womb. You see light.

PRIBRAM: What happens ontologically is that somehow as you go on through those five years you distribute the thing, get it split up into parts all over the place, and then, what seems to occur, is that some new constellation, new way of getting it put together again, occurs at that moment of enlightenment. It's some process of that sort.

MAN: After the five years, what happens?

PRIBRAM: I don't know, I just got it to that stage.

SPENCER-BROWN: Let's simply go through the procedure again. The covenant with the world that the child rapidly has to make is "Right. I am not allowed to notice this." But the child perceives where the lines are drawn and not drawn, and then suddenly it realizes that it must put on the same blocks, otherwise it will not be accepted. There is a moment of sanity when this happens. However, it's "good-bye" for quite a long time, I don't know how long. If it is to survive, it's "good-bye," and "hello"—"Hello, world."And now instead of it being able to deduct, because it sees that is fully outlawed, now it goes through the game of those people who know best, and who are teaching it; and in order that they can have the game, and it can play it, it must pretend to know nothing, so that they can now pretend that they are bringing it up and educating it. And so it then has the—It goes through the learner stage of playing the game, of looking at things and being surprised: "Oh, look at that!" "What's this for?" and so on. And thus, the whole proceeding of playing the game that there is an objective world, which you can run around and look at and pick flowers and bring them back and say, "Look." It's when one gets very far in this game and begins to wonder what it's about and how it is that we do find something outside, and it does appear to have some structure, and so forth, and to come back and base it on what we are doing, that we begin to see—that we begin to ask the question, "Well, what is there outside?" We begin to realize that what is outside depends on what is inside.

Why We See the Same Things

SPENCER-BROWN: One of the questions that we might ask is why we appear to see the same things. Why does it appear—I can see the moon and you can see the moon. If you are a different shape from me, then you should see a different thing. Well, if you take it back far enough, we have this;[14] in other words, from one mark, we have any number of identical marks. This is a process here. Actually, mathematically, this arrangement is still only one distinction. It has essentially the same rules as this one. Just let's have a look and see where you are. Outside, outside, inside, inside, making one crossing.[15] Outside, outside, inside, inside, outside, outside. There is no difference mathematically between that and that. You make the same number of crossings, you get the same thing. This is the inside and this is the outside. In other words, we can form another illustration of, that is the same as this, in the point where they condense;

---

14. As in ⌐ = ⌐⌐⌐ (Fifth Canon, Expansion of Reference, confirmation, LOF, p. 10).
15. Actually, the text describes zero net crossings. Spencer-Brown may not have spoken the name of all the crossings in his diagram on the chalkboard; and the diagram itself was nowhere recorded.

but even so, we can make it look like two. Insofar as you and I see the same moon, we do so because it is an illusion that we are separate. We are the same being. We only appear separate for the convenience of filling space. Of course, we can't have empty space-we'll have to fill it up with something.

WATTS: Parkinson's law.

SPENCER-BROWN: Yes. And with only a limited material to fill it up with. So since space is only a pretense, the observer, in filling space, undergoes the pretense of multiplying himself, or stationing himself. But two people are only like two eyes in one of them. The scientific universe, the objective form which we examine with telescopes and microscopes, and talk about scientifically, is not the form which our individual differences distinguish, It's the scientific, objective universe observed with the part of us that is identical for each of us. Hence its apparent objectivity.

What is called *objective* in science is where we actually use our individual differences, where we say, "Well, that's rather different from that," if, in fact, what we observe depends upon that, and so forth; therefore, that's not an objective distinction, that is something which is a personal view. And that is not what science is about.

WATTS: Well, how would you react to the remark that what you have been saying is a system that used to be called *subjective idealism*, in which you have substituted the structure of the nervous system for the concept of mind?

SPENCER-BROWN: Well, I can go along with the nervous system, because the nervous system is an objective thing in science as well as a thing we observe—as the constants of what is called a body, which is an extension into hypothetical space of a hypothetical object. I have never had this thing about brain at all. "Inside my something brain"—"my teeming brain." I have never felt that my brain is particularly important.

WATTS: Are we talking about the structure of the sense organs?

SPENCER-BROWN: Yes, only to bring us back to the fact that we have made the distinction between the world and ourselves. I have played the science game to show that even in science, playing the science game, which is to say, "Right. The reality is thus: There is a distinct *me* with senses. There is an objective world with objects and lights and things flashing about, and when I see that window there it means because there is light coming through that window focused through the lens of my eye on my retina in a certain pattern, which goes through the nervous channels to the visual area of the brain, where it all project into a muddled, upside-down..." and so on, with the whole scientific story. And the trouble with the whole scientific story is that it leaves us no farther, it leaves us no wiser than we were before. Because nowhere does it say, "And here, this is *why* that is how it appears."

But if you play that game, as I was doing for the purpose of illustration, one still finds that, operating philosophically, and saying "Suppose I change all my sensory forms?" Now, the whole universe is changed, I am only doing this to show that even playing the science game, whatever game we play, must leave us the same place. Even playing the science game, we see that there is no distinction between us and the objective world, except one which we are pleased to make.

KELLEY: Can you tell us something about the Fibonacci development?

SPENCER-BROWN: Since not everybody here has mathematical training, it is something which, if I have the breath, I will do later with you and perhaps a number of other mathematicians. To explain it to the people—beautiful idea mathematically—it does involve some rather lengthy exposition.

Gödel's Theorem: Completeness and Consistency

KELLEY: Maybe another question that wouldn't be too far out: what's the definition of the *accidental factorization*?

SPENCER-BROWN: I was hoping you weren't going to ask me that. It has been a long time since I've done this. Again, it is something that I would—

KELLEY: As I recall, I think that Gödel's theorem basically says that in an algebra, you don't have completeness and consistency.

SPENCER-BROWN: Not in *an* algebra. In the common algebra of numbers you can't have both. This is where so many people go wrong over it.

KELLEY: But the result is no more general than from the common algebra of numbers?

SPENCER-BROWN: No. For example, in this algebra you do have completeness. I have proved it. And consistence. I have proved both. In *Laws of Form*, you find proven consistency, and, in Theorem 17, proof of completeness.

KELLEY: OK, now, does Gödel's theorem only apply to the algebra based on real numbers? That is, it's beyond the integral, it's a field, right? The field of real numbers, where you've got multiplication, addition, and associativity.

VON MEIER: A system at least as complex as arithmetic.

KELLEY: Well, I am trying to find out where the boundaries are.

SPENCER-BROWN: It does, in fact, have interesting boundaries. This is a very common error among mathematically trained people, that it does, in fact, apply all over. It doesn't…it's not applicable to the primary algebra, which is both consistent and complete. It is not applicable in a modulus where algebraic factorizations are the only factorizations. Gödel's theorem doesn't apply. The modulus is both consistent and complete.

The ordinary algebra of number, not introducing the complete system—for example, the algebra of the positive and negative integers—now Gödel's theorem applies, provided you use both constants, multiplication and addition. The difficulty is that you have got two voids. You have got a void of zero in addition and you have got a void of one in multiplication. The constant you put in makes no difference. Interestingly enough, it doesn't apply to the complete number system.

KELLEY: It applies in an integral domain, but it doesn't apply in the field.

SPENCER-BROWN: In the whole, in the complete field, using real and imaginary number, no. Complex numbers. It doesn't apply.

KELLEY: OK, what if you have just the field of real numbers, not including complex numbers?

SPENCER-BROWN: Then I think it applies.

KELLEY: Now I think I am beginning to get an idea of where the boundary is: When the field Just includes real numbers, it applies; but when you have complex numbers, it doesn't.

SPENCER-BROWN: It happens that way, yes. But it is beautifully illustrated in cases where you are working with a modulus where you have no accidental factorizations. If you find the algebraic factorizations, you have found them all. Whereas when you are not working with this modulus, when you are working with the integers, when you have found the algebraic factorizations, you still haven't found them all. The others are called the accidental factorizations.

End of Session Two.

AUM Conference: Session Three
Tuesday Morning, March 20, 1973

transcript created and edited by Cliff Barney
from recordings by Kurt von Meier

SPENCER-BROWN: I was asked if I would deal with certain technical points yesterday. I said that I would, but I feel that it would not be in order for me to do so because it does seem *[like]* talking to a very few people—just one or two—who are interested in specific, narrow questions of research; and I do feel that this is more in order if, in fact, the audience was of the same kind of people and then we could talk about whatever technicality may be in order and required.

Also, in respect to certain technicalities, I do like to prepare myself for them. I once went before an audience during a course of 20 lectures, and I had to prove a theorem which I thought I knew so well I didn't need notes; but I started on the proof and I couldn't remember it, and so I asked for help from the audience and they were helpless. And I spent the whole lecture trying to find the proof, and they were trying to help me, and in the end it was wasted. That was the only time I have ever done that, and as soon as I got back home, of course, I remembered the proof.

I don't think any of the technical points are as technical as that, but I do like to be prepared and I do like to give a technical exposition to a wholly technical audience; because, if there are others who are not specifically interested in the technical question or haven't sufficient training to understand it, I think it would be a little unfair to them to have to listen to something which doesn't mean very much as far as they are concerned—even though they may be very obvious points to someone with mathematical training.

There was a mathematical lecturer, a professor at Cambridge in my college, Trinity, who was giving a lecture...and he was just coming to the end, he was just rounding off and saying "It is obvious that—" "But sir, I don't see that it is obvious." So he had a look at the blackboard formulae and did a few calculations, and time for the lecture was finished and everyone got up and went away. And this student who had asked the question still stayed on. And he tried something else, and then said, "Excuse me just one moment, I must go back to my room and look up some books. And so he went back to his room, And then five hours later he walked back into the lecture room and there was the student, still waiting. And he got up triumphantly onto the platform and said, "Yes, it *is* obvious." That is what is obvious in mathematics—the more obvious it is, the longer it takes to find it.

No **Not** *Sense*

BATESON: To take off from yesterday's turtle, somewhat—

SPENCER-BROWN: The turtle, the tortoise—oh, yes.

BATESON: My interest, if there is anybody who will go along with it—if it's a nuisance to them, would they say so—is in, amongst other things, animal communications. And what goes on between animals is evidently characterized by, amongst other things, the absence of *not*—the absence of a simple negative. While they can forbid each other—say "don't—they can in general not deny a message which they themselves have emitted. They cannot negate,

Now, the messages which they emit tend to go in the form of intentional groups, or something which is part action, and part stands as a name for the whole, in some sense. So their showing of a fang is a mentioning of battle. Not necessarily the beginning of a battle; possibly a challenge, possibly a mentioning with a question mark—I mean, "Are we here to fight each other?"

It's sort of in the hope, that I am here, that your laws of form calculus might be the sense on which to map this sort of sound. We have a two-legged language which is very unsuitable for mapping what goes on between animals. Indeed, it is unsuitable for mapping what goes on between people.

SPENCER-BROWN: Before I answer that, I should have to explain that Prof. Bateson has written most lucidly on this theme, particularly in a little metalog, in the form of a duolog, isn't it, between a father and his daughter?

BATESON: It's merely a dialog, yes.

SPENCER-BROWN: —about instinct and about the language of animals compared with the language of us. This is, I believe, now published, and the title is *Steps to an Ecology of Mind*.

There is this delightful little duolog in that book. When I first read it, some years ago in London, I found it contained very profound observations on communication and—excluding, really, in terms of animals like whales and dolphins which do seem to have a form of communication, which, if you could divine it—I hope I am right, John—is at least as efficient as ours and probably something like it and maybe better. I think that this causes something—possibly the same problems that they have. Although they may have something superior to laws of form, in fact, having got something that is more important, or more fundamental, than *not*. Laws of form comes effectively from the licensing of the *not* operator in logic. What is of interest in Gregory Bateson's account of the animals is that they don't so much communicate as commune with us and with each other. And I would like to make this distinction.

Communication and Communion

SPENCER-BROWN: Amongst the other distinctions that are not commonly made, or, if they are made, are not made consistently, is the distinction between communication and communion. Communication happens according to physical existence, in some physically detachable sense, and the characteristic of communication is that what goes on goes on at the same level. One can take at the level of physical existence, nervous events ordered by sound waves. For example, wireless waves, or what have you, all detectable in physical existence, followed by a perceptor of information, et cetera, et cetera, et cetera.

Now, it is my thesis that communication is superficial to communion, and without communion, there is no communication, really, at all. That is to say, if there were no communion, which I will now define as a fitting on another level between the communicants—if there is no communion, as indeed there sometimes is not, then what is communicated, when it reaches the other end, is not understood.

The more perfect the fit on the communion level, the less needs to be communicated, the more that can be crossed from one being to another in fewer actual communicated acts. In *Laws of Form*, this is expressed in these two laws, or at least there are pictures of it in the two laws early on, in the canon of contraction of reference,[1] whereby, as people get to know each other better—a gang of kids go about and one word or even half a word is used to express a whole community between them. Whereas when people do not know each other, this has to be expressed in a whole book. But between people who do know one another, however, there is no need for a book, it can all go in half a syllable.

Now when one is communicating, for example, with one's cat, that doesn't have the sort of language we have, or if it does, we don't know it, then it is done in this kind of way. It is done because you know each other. And when my cat says "Meow," I sometimes say, "What do you mean, 'Meow'?" But this is a game, because if I consider it, there is never a time when my cat says "Meow" that I don't know exactly what he means. Why I sometimes say, "What do you mean, 'Meow'?" is because I can't be bothered to get up and give it the fish or open the door or pet it. If I am honest with myself, there is never any doubt whatever. Although it says "Meow," it makes it quite plain to me, by the context in which it says it, exactly what it means. And if I pretend that I don't understand, it knows perfectly well that I am being awkward.

So, to put it on the positive side, if one doesn't make this pretending game and say "Really, the cat ought to be talking like we are," but goes on the level to how it can respond, the communication between a man and the animal can be so complete as to be almost unbelievable. The understanding can be very much greater than between two human beings.

Now, with this question of how is it—I am going a little beyond what Prof. Bateson says in his duolog, where he raises this point. The question of how people get into fights, when, in fact, this is a mistake, they got into one by mistake, through one or the other—people or animals taking what... You see, for example, if I tease my cat

---

1. Second Canon, p. 8, *Laws of Form*: "In general, let injunctions be contracted to any degree to which they can still be followed." Although Spencer-Brown does not mention it specifically, the other canon referred to may be No.4, the hypothesis of simplification (p. 9): "Suppose the value of an arrangement to be the value of a simple expression to which, by taking steps, it can be changed."

and it begins to think "This is enough," then it comes round and gives me a little nip. Now this is not nearly as hard as it can do it. The nip is the same, when it is a warning nip, as when it is a completely playing nip. And where I have seen things go wrong, then—to get on the subject of where things go wrong—you may have an entirely neurotic animal who does not distinguish between the gradations of nip. Because when an animal has been made neurotic, what it's lost is its capacity to distinguish. And what has happened in its place, it's been devastated in some way; and it either is completely anesthetized to what is going on, or if it perceives it, it perceives it fully. It perceives a nip of a certain strength as complete war.

BATESON: It's not a problem of your initial thing and the token of it?

SPENCER-BROWN: Well, I am going to that.

I am trying to treat it, first of all, getting into the open, as you are doing with metalogic—getting into the form of extremely simplified and yet extremely sensitive communication of animals. The cat has not a great many modulations of its voice and still fewer twistings of its tongue to make what comes out different.

WOMAN: It has the widest range of sounds of any animal.[2]

SPENCER-BROWN: It has a wide range, yes, but it doesn't have words like we do. For a lot of things, it says the same thing, but in a different context, looking a different way, or what have you, which can mean in one case "Play with me," in another case "Feed me," in another case "Open the door for me," and so on. Now it does not have any problem with other cats unless they are neurotic, unless they have been in some sense devastated, in which case it may get into a fight mistakenly. And it has more difficulty with humans, because humans tend to be more neurotic. But it doesn't have the problem with a human being who understands the gradations the cat does, and is sensitive to them.

Now, having gone that far, let us consider something which Gregory Bateson posits, and I tend to agree with him: The one thing that a human being has in his language, which other animals, if they have a similar language, don't yet have is a word or an expression having the effect of *not*. Now just as human flesh can accommodate cuts and bruises better than burns—it doesn't seem to know that so well—so the human mind can accommodate to positive sentences much better than to the same sentence with *not* stuck on there somewhere. *Not* appears to be a recent acquisition in language. In fact, if this is so, it would be that we were least adapted to it, most unreliable with it, and we do agree that we—indeed, it is well known in business when one has to get something done, that you have to be very careful to put what you want doing in positive terms. Don't put it...like I'm putting it.

---

2. Excepting, probably, birds.

My professor of anatomy, J. D. Boyd, didn't appear to understand this. Because he was a very good lecturer—he had if anything one fault. When he was describing some part of the human anatomy, he would make a list always of the common mistakes that students made as to where a nerve went, of whatever it may be, you see. It doesn't go there, he would always write, and it doesn't go there, and this doesn't happen and that doesn't happen like that. And then he would—this would come out in his lectures and he would say "I cannot understand this," he would say, "I told my students exactly the mistakes they should avoid, and these are the very mistakes they always make."

LILLY: They were following directions.

SPENCER-BROWN: They were following directions. And whether the directions have *not* tacked on somewhere or not, is something which they forget. And indeed, this is so obvious that there are ways of maligning people—for example, a picture of somebody in the paper and the caption underneath—"Denies Cuddling Policewoman."

Or one could even go further to the well-known joke of the king who wanted to be able to turn lead into gold; and who—he put an advertisement in the local paper for a magician who could do this. And the conditions were that if the recipe failed, the magician would have his head cut off. Well, lots of magicians came for the pleasure of having their heads cut off—there is one born every minute. And finally a very good magician came and—well, he would get oil and bring it to a boil, and put in a toad's liver, as an experiment; then you put your lead in and count to 15 and then you add a pinch of salt. You do all these things, you see—this, that, and the other, and so on.

Having finished the recipe—the king was writing it down—he was just about to be taken off to where he would have his head cut off if the thing doesn't work. And just as he is about to be taken off, he says "One moment, your Majesty, one moment. There are just two more instructions which are necessary to this recipe. One is that it must be done by your Majesty himself—you may not delegate. And one more thing, your Majesty, one more thing, you must not think of a hippopotamus while you are doing this."

He kept his head. We are least adapted to *not*. *Not* is the worst order to give anybody, the most confusing order, and the most unlikely to be carried out properly. I do think that, apart from possible animals who have a language as evolved as ours, I do think that it does make for a very different way of seeing the world; or, to put it more accurately, it does make for a very different world. The world waxes or wanes as a whole. The world of the happy is totally different from the world of the unhappy.[3]

Manifesting the Form

SPENCER-BROWN: So one can either say, "there are various ways of seeing the world, or one could say, "There are various worlds," which means the same thing.

---

3. Wittgenstein, Tractatus 6.43

How could there be a difference between these two? As soon as we have *not*, we have a kind of world that no animal without *not* ever sees. And since, in *Laws of Form*, the laws of form can be described as coming from granting a license to *not*, it is, therefore, this universe of the *not*-speaking animal that this particular form is about. The form itself manifests in as many ways as there are ways of distinction. As in the *Tao Te Ching*, we start with the first proposition, "The Way, as told in this book, is not the eternal Way, which may not be told."[4] The eternal way may not be told because it is not susceptible to telling. It is too real for that. It manifests in as many different ways or different expressions as there are differences in the beings to which it manifests. So that *[when]* I speak of *the form*, that is never the form that is spoken. The form which is spoken is the form as it is manifest to us, as the particular beings we are, with our particular *not* culture, our particular *not* language, and our particular conventions of life.

And when one looks at a cow in a field and somebody says "What's it doing?" well, I say, "Well, I think it is contemplating reality." And they say "Don't be ridiculous, how can a cow contemplate reality?" "Why not?" I say. "What else does it have to do all day? What else has it to do? The thing is contemplating reality, what else could it be doing?" But the form as it is apparent to a cow—although it is the same form, it is the Way without a name—how it manifests to a cow is not how it manifests to me. How it is expressed to a cow is not how it is expressed to me.

BATESON: Could one have identified self, without a not?

SPENCER-BROWN: Well, that's where you return to the tortoise—because of the game we play, where we have defined there is a me inside—my body and the world outside, and we don't even wink when we are doing it. We take it dead seriously. And what we have, you see, to make all this so dead serious, is we take so dead seriously the *not* boundary. And to us the form of the fiction is a boundary with *not*—not one side or not the other.

Now to recapitulate, how of course can there be any space, where would there be for it to be? How of course can there be any time, when would it exist? The world being the appearance of what would appear if it could, if the impossible were able to come about. Now if the impossible comes about, or appears to come about, in as many different ways as it can, according to the form. And in this particular existence, we have the privilege, if you put it that way, the privilege of actually viewing from the apparent outside, other points of view, like tortoises, which are other ways in which the impossible would manifest if it could.

MAN: Do you distinguish between *appearance* and *is*?

SPENCER-BROWN: Not at the moment. I would do it if it was needed.

---

4. This is the epigraph expressed by the Chinese characters opposite p. 1 of *Laws of Form*.

MAN: The reason I ask is that to me the primitive is not *not* but *is*, and the distinction between animal communication—and I got this from Gregory, standing on his shoulders as it were, looking either down or up, depending on how you interpret my interpretation—it seems to me that the *is*, the "isness" of communication is what is particularly human. An animal just—

VON MEIER: No, it's only peculiar to a language. Russian has no copula. Chinese doesn't use the verb *to be*—doesn't articulate being with a special verb in the language. It sets things bedside one another, which is a sense of the Greek word *paradigm*.

WATTS: Chinese indicates *is* with *that*.

VON MEIER: To translate English?

MAN: To me, I can distinguish between just pointing, saying "Lois," and saying "That is Lois; she is Lois."

WATTS: There's a statement in Buddhist literature, "Void is form." Now the *is* word is not our *is* word. It's "Void that form."

Being and Existence

SPENCER-BROWN: Well, one must distinguish between *being* and *existence*, *being* being deeper than *existence*. *Existence* is less important than *being*. However, even *being* is not the most important. As to *existence*, well, there is a whole world that *be*, which don't even *exist*, and the world that don't exist is far more real than the world that do.[5]

We have an astronomer who talks on the television, and he answers questions—he gives a monthly program and then he also reads his letters. And the letters are usually, "Well, what happens in the center of the sun?" Or, "Is the Andromeda nebula a spiral? What colors come out of it?" and so on and so forth. And he was answering the questions in one program and the final question was from a lady who asked on a postcard, asked a short question: "What I would like to know is none of these specific questions—what I would like to know is, why the universe exists at all? And he put on his most Satanic expression, and just before the fade out he replied, "Does it?"

Intent of a Signal: What Is Not Allowed Is Forbidden

WATTS: Would you reflect briefly on the word *not* in the context "Whatever is not expressly permitted is forbidden"?

---

5. See p.101, Laws of Form. "to experience the world clearly, we must abandon existence to truth, truth to indication, indication to form, and form to void." See also the discussion of the five eternal levels, near the beginning of Session Four, under the heading "Five Eternal Levels & the Generation of Time."

SPENCER-BROWN: You mean, "What is not allowed is forbidden?"[6]

WATTS: Yes.

SPENCER-BROWN: Well...this is the form of all documents that have to be precise. And mathematical and legal documents are the same in this respect. The point is that you cannot be precise in the expression of anything at all unless you make this rule. How otherwise could one, you see—because one would never know, if you didn't expressly allow it, what was allowed. If you let any "allows" slip through the gate, now you cannot be precise.

The reason we don't realize in ordinary speech or ordinary communications that this is the law of precision is that we have so many unspoken conventions, which in the same society are the same for the same people. That is why, when somebody is playing a game and they suddenly realize that something new that nobody has ever done before in this game is in fact permitted by the rules, and they do it, there is a cry of "Unfair," "Taking advantage of the rules," and so on. And then sometimes the rules are changed. Since it is required that it *[be]* absolutely precise what may or may not be done, there must be this rule that what can be done is what is specifically allowed.

CLIFF BARNEY: Can you turn it around and say whatever isn't forbidden is allowed? That was the rule in the Garden of Eden.

SPENCER-BROWN: Yes, you can do that. What is not forbidden is allowed. Because whenever you have one law, in the next level of existence you have a reflection of it. And, in that sense, you're not talking of mathematics. In mathematics, which has to speak precisely, this is the first canon. I don't actually know that it has been expressed before in any mathematical document.

WATTS: It has been expressed politically.

SPENCER-BROWN: I am talking mathematically, Yes, it may have been expressed politically. It is the most commonly broken law when people come up with false proofs; a mathematician will immediately recognize that they have done something which they haven't in their rules actually allowed. And this is how anything can be proved. It's known and it is immediately recognized that the canon is broken in mathematics; but as far as I know, it has not been expressed before as a mathematical canon.

VON MEIER: Identity elements in a group?

---

6. *Laws of Form*, p.3: First canon, Convention of Intention.

SPENCER-BROWN: No.

VON MEIER: In number series, zero?

SPENCER-BROWN: No, no, this has nothing to do with the elements. This is a canon of how you begin to make your laws. And the first canon is that while you are making the system of the laws of instruction, unless you allow something, then you forbid it.

LILLY: In other words, no covert contracts.

SPENCER-BROWN: Yes, no covert contracts. It's all overt, this is the game in mathematics. The whole thing is overt. That is why, whatever we do, we must allow.

WOMAN: Does this cover right or wrong?

SPENCER-BROWN: No, this is mathematical. We are not anywhere near right or wrong, you see. The mathematician who is used to the fact that we have, in fact—well, it began with the covert convention, that became overt, that we are only allowed, in defining *operator*, to define it as operating on two variables. That's what gets us into such trouble, you know. The Sheffer stroke, for example, it is not allowed on more than two and it is not allowed on less than two. So *not A*, with a Sheffer stroke, must be done A stroke A. In fact, if you will read the first few chapters of *Laws of Form*, we specifically allow the operation on more than one variable. Since we have allowed it, we may do it. And it is not relevant to refer to the forbidding of it in other calculi.

This is the difficulty of reading mathematics; one has to be able to read just what it says, because there is nothing in it that one may assume, apart from the knowledge of the language used and how to count—these are the only things taken as common.

End of Session Three.

AUM Conference: Session Four
Tuesday Afternoon, March 20, 1973

transcript created and edited by Cliff Barney
from recordings by Kurt von Meier

Last Requests

SPENCER-BROWN:
Well, on this last occasion I have received many requests to speak about different things, so perhaps I should deal with the requests in order. The first request is for some guidance on how I use the principles in the book *Laws of Form* in everyday life. And, since one of the principles which is exemplified in *Laws of Form* is that you don't tell how it is that you use the principles in laws of form in everyday life, I cannot comply with this request.

I have had another request to talk about the relationship, or distinction, call it what you will, between male and female. I guess you are asking the wrong director of the company that I represent. It is my fellow-director who is always going off on courses about how to get along with the opposite sex; while I have to stay home and actually do the work. So I get so little time—I have so little experience of either sex that I don't feel that I am qualified to say anything, apart from the fact that, even if I could say something, the subject is so big that it would seem a pity to start it on what must be the good-bye occasion. I can only say, that the more I have to deal with the opposite sex, the less I think I know about either myself or themselves, or vice versa.

In other words, I think both of these subjects are, to some extent, personal and private talks, and other games, rather than for a relatively public lecture, for which I don't feel I have the experience to talk in any way which I feel would be authoritative.

I do feel there's little left now for me to say except to thank you all very much for listening and for being such an extremely good audience as to prompt me with what to say for, some, how long is it, two, four, seven-maybe nine, ten, twelve hours, since I have been here. I feel I ought to be making a speech. But, all good things come to an end, and I could I suppose answer maybe two questions, if they don't fall within the limits of what is unanswerable. If anybody has any last request.

Five Eternal Levels & the Generation of Time

LILLY: I have one.

Footnote One in *Only Two Can Play This Game*. You say "To cut a long story short, it turns out that there are five orders, or levels, of eternity."[1] Would you diagram those for me?

SPENCER-BROWN: Diagram them?

LILLY: Put five lines and a label at the end of each line. Starting with zero. I find that the following discussion gets a little unclear, because sometimes you are going in and sometimes you are coming out, And I am not clear where you are when you are doing this.

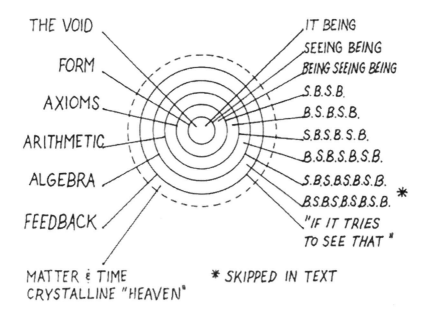

Reproduction of Spencer-Brown's drawing on board by Cliff Barney.

SPENCER-BROWN: Well, it looks like an electron. Indeed, in some circles it *is* an electron. Level ought. Level one. Level two. Level Three. Level Four. There is a diagram of the five orders to eternity, the five levels. They are brought about—there is no about until they are brought so—by it being, it seeing *[being]*, it being seeing being, it seeing being seeing being, it being seeing being seeing being, it seeing being seeing being seeing being, it being seeing being seeing being seeing being, and it finally seeing—it seeing being seeing being seeing being, seeing being—and if it tries to see that,[2] it finds it can't without going half blind and coming out into time.

---

1. Keys, J. (1972). *Only Two Can Play This Game* (p. 126). New York: Julian Press.
2. Spencer-Brown missed one level in this account. After the "it seeing being seeing being seeing being seeing being" sequence comes "it being seeing being seeing being seeing being seeing being, and *then* the phrase "and if it tries to see that…"

LILLY: So time appears where?

SPENCER-BROWN: Next one. The first time.

LILLY: The first time appears outside the four levels.[3]

SPENCER-BROWN: That's the next look it takes, but it finds it can't see that without going half-blind. After all, as I say there, after all, time is a one-way blindness, the blind side being called the future.

LILLY: Where's *flippety*?

SPENCER-BROWN: Well, it corresponds laws of form to the void, the form, the axioms which see the form. You have to get this number right, you see; because it is the number that Dionysius counts on his orders of angels, but he doesn't always arrive at the same answer. Then you get the arithmetic, which is seeing what becomes of the axioms. And then you be it to do it, and in being and doing it, you find that, being and doing, you see the generalities of it, and that is the algebra. And while you are seeing you notice you have got equations, something equals something else, and then suddenly you decide—aha, supposing what it equals goes back into what it comes from? Now you have generated time and matter all at once; There can be no matter without time. Time and matter come simultaneously. But this is the first matter in which the orders are counted, and it's called the *crystalline heaven*,[4] but it is not, really. Technically speaking, it's not really a heaven. And as it keeps recompounding, and re-inserting, it gets the appearance of being more and more solid, until it really, you know, is pretty durable.

LILLY: It can kill you.

SPENCER-BROWN: Well, that may be.

VON MEIER: It will sustain our life.

SPENCER-BROWN: At the grave, you begin to wonder, Just who is there to be born, to be duped, to be killed? Just where is there for it to be, and just when is there for it to happen? Or, as some sage said, when he was dying and somebody was crying, and he

---

3. Lilly means five levels.
4. Merriam-Webster defines *crystalline heaven* as "either of two transparent spheres imagined in the Ptolemaic system of astronomy to exist between the region of the fixed stars and the primum mobile in order to explain certain observed movements of the heavenly bodies." The primum mobile, in the Ptolemaic system, was an outer sphere supposed to move around the earth in 24 hours, carrying the inner spheres with it. So the crystalline heaven was outside the primum mobile—apparently Ptolemy's *fifth crossing*.

said "Why are you crying?" and he said "Because you are leaving us." "Just where did you think I could go to?"

Consciousness

LILLY: Where does consciousness first appear in that setup?

SPENCER-BROWN: Well, it's there all... What we consider to be consciousness, in the sense of... You see, it's not called *consciousness* until suddenly you have names to begin with. But there is no meaning. It is co-extensive with existence, because what could it possibly be—anything be, let alone exist—without its being a form of consciousness of its existence. There is no problem of consciousness, none whatever. Its meaning is coextensive with whatever there is.

WATTS:
> There was a young man who said, though
> It seems that I know that I know,
> What I would like to see
> Is the "I" that knows me
> When I know that I know that I know.

I think that's what you've diagrammed.

Waves and Particles

SPENCER-BROWN: In the construction of matter, all that happens is that we create the temporal and the material together by imagining that the outside feeds back into the inside. We then have a succession like this of marked and unmarked states generated by this.[5] as long as the tunnel is there, this goes on, and when the tunnel disappears, this is of a particular length, a wave train. But, if this[6] were to go past, similarly, this would also appear as a wave train, and yet, as it is, it is—it could be an electron. If there is only one electron left—you know, all the rest were done away with—it would be quite sufficient to recreate everything. So long as there were one bit of space left, it would be quite sufficient. All would grow out of it. If there were nothing left at all, it would be quite sufficient.

If It Is. It Isn't

BRENDAN O'REGAN: One wonders how there can be mathematical theorems which exist about space which does not exist.

SPENCER-BROWN: Who said space did not exist?

---

5. See p. 60, *Laws of Form* (Fig. 1 and Oscillator Function) in 1972 Julian Press edition.
6. See p. 61, *Laws of Form*, Fig. 2. in 1972 Julian Press edition.

O'REGAN: You did yesterday. I was just following you through.

SPENCER-BROWN: Oh no, no, I said, you see, in the common usage of existence, space only exists. On the other hand, if we go deeper, go to another level, and say "What does existence consist of?" then we can produce these semiparadoxical statements that say "Well, it is what would appear if it could." This leaves it open as to whether it has or hasn't. It doesn't go one side or the other of the boundary. It leaves you still in the form, at the point of indifference.

It is so difficult, in the Western teaching, not to plug for one team or the other—to think that one must make a choice between either and or. In reality, it is neither one thing nor the other. There is no need of this choice. It neither is whatever we say it is, nor is it nothing. It neither exists nor does not exist. Because, remember, we have created it out of what is, in the Russellean paradox, the forbidden contradiction. It has been created out of "If it is, it isn't; if it isn't, it is."

And this is why, to get back to the reality, we have to undo this. We do see it precisely because it neither is nor isn't whatever we see it as. Because if it is, it isn't, and if it isn't, it is, and that is why we see it as a material.

O'REGAN: This point that Karl Pribram was making about that with our abilities to perceive from the sensorial point of view—One might argue that we can only perceive difference, and, in a certain sense, if you say that we can only see it because it is or because it isn't, is it the process of it becoming and not becoming that we perceive?

SPENCER-BROWN: Yes. Hence, once you get to this stage, where you are once in time, now everything is a vibration of it. Vibrations—as we know, the mathematics of vibrations is always the equations with the imaginary value—if it is, it isn't, if it isn't it is. Whichever it is, it isn't.

VON MEIER: It seems like the inverted image when we see with our eyes, corresponding to our tactile knowing that it is not upside down; so it's an internal systems check.

SPENCER-BROWN: I am not sure that is on the same level.

VON MEIER: We have two aspects of reality to deal with—our tactile sense of gravity, knowing something, a pyramid, to be like that; but, nevertheless, seeing it and then having to translate it in our brains. We have to go through that redundancy step.

Understanding and Standing Under

SPENCER-BROWN: What has to be learned in any understanding is that one can stay at the same level—one of these levels—or the others as we get on, but there is no understanding by making... Say, here is the level of physical existence, with all the light waves and solid objects, and so on. They are not really very solid. You know, when you get down to trying to see them, they disappear. It is the illusion of solidity.

If as much of the science game, in certain aspects of it, goes and says, "Right; well, we explain that in terms of this," everything at the same level, there is no understanding. Because *understanding* means literally what it says: You go into another level and stand under. And this is what we[7] are forbidden to do. It takes a long time of relearning, to go from level to level. When you are talking in one level, what is described is quite different from when you go to another level; and, having translated down to another level, we don't have language that will enable us to do this. And that's why when we talk with understanding, it sounds to people at the same level all the time, it sounds like nonsense. They say: "You are contradicting yourself." Of course you are contradicting yourself, because what is at this level is an image. It is all reversed.

Contradictions

MAN: Do you shoot back and forth?

SPENCER-BROWN: Yes. That is why all the mystic utterances contradict themselves. Wittgenstein pointed out that a measure of a tautology, a statement which is true by the very nature of its form—"If A, then B and A, therefore B," that's a tautology—a form of words which has the same truth value as being true whatever you substitute for the variables—Wittgenstein pointed out in *Tractatus* that all tautologies say the same thing, i.e., nothing. They say not a thing. What he missed out was that—he missed out the image of this—he missed out the other end of this continuum, the other end being the contradiction, which says everything. You can't say all about it without contradicting yourself.

We have so many social values that spill over into our university training, even in so-called objective subjects like logic. Somehow, contradictions are good—sorry, somehow tautologies are good and contradictions are bad. Now this is childish, childish prating, and you can see how it has arisen. It comes from the nursery, as do most of these things. The nurse says, "Naughty Johnny, you have told an untruth." ... "Good Johnny, here is a sweet—you have told me the truth." Since tautologies are true, and contradictions are untrue, technically speaking, we have carried this over—contradictions are naughty and tautologies are nice, good things. So one of the reasons for the whole cultural forbidding of mysticism is that it deals in statements that say everything and, therefore, must be contradictory, therefore must be logically false, and, therefore, are naughty.

WATTS: A contradiction is a no-no. We've become used to that expression in the United States.

---

7. Scientists.

SPENCER-BROWN: Well, I don't know what that means. I am going to come back to one of the beautiful things of Roth, you see. As I called it in *Only Two*, the spectacular introduction to *Dionysius the Areopagite*. It really is spectacular. I do recommend it. It is much better than Dionysius. It is much better than the book. He originally has this marvelous thing which we were talking about earlier: "and all this went on in perfect harmony until the time came, for time to begin." Utterly contradictory, but, you know, it's the only way to talk of this, because we have to talk in language which—language, you see, is built for a level. That's why when you learn a language, you know, you are confronted with such fatuities as "The pen of my aunt is in the posterior, whereas my…"; you know that sort of thing. It's all on this level because this is what makes it respectable. Language is not something designed for shifting gears up and down the levels.

Injunctive Language

LILLY: You talk about the injunctive use of language, however.

SPENCER-BROWN: Yes. This is the only way we can do it,[8] because it has to be done in mathematics, and also has to be done in the tutelage of any discipline. The descriptive use of language just describes, you know. We say "describe a circle," and here we have described it, you see. The injunctive use of language now enables us to cross…cross the line.

Injunctive language has to be used in any field in which the discipline itself is to move from level to level; and this is why the whole of mathematics, which is simply about this—apart from the precision and description, which is an art in itself, taken at one level, and this is why the language of mathematics is so beautiful—but apart from that it is nothing but orders: do this; stand there; consider that; observe this; move here; call that over there; mix these two.

LILLY: Once you have absorbed the cookbook for changing levels, do you need it any more?

SPENCER-BROWN: Once you have observed the what?

LILLY: Absorbed. Once you have taken all the injunctions, the list, your set of instructions, and absorbed it, and now it's part of your thinking machinery, yourself, do you need it any more?

SPENCER-BROWN: Only—Well, it's like saying, "Do you need it if you want to play a piece on the piano?" "Do you need it if you want to read a bit of mathematics?" You

---

8. Except perhaps by a trick: Latin *pungere*, to prick, gives *puncture* and *pun*; the pun pokes a hole through the boundary.

need the experience of being able to read injunctions. You see, most people cannot read injunctions. One of the things one has to learn is to read injunctions.

LILLY: —And use them.

SPENCER-BROWN: And follow them, yes.

LILLY: A cookbook can only be used by a cook.

SPENCER-BROWN: Well, a cook has the experience. If it is used by a non-cook—a non-cook has to take more care. It can be. Just as mathematics can be used, you know, by somebody who has never seen it before. But the care has to be very great. In a cookbook, you know, the recipe only lasts a few pages, whereas in mathematics, a complete argument may last 150 pages, maybe, and that means that every step previous, in order to be able to follow what's going on, has to be remembered. Otherwise you lose track of what you are doing. And part of the discipline is learning to remember.

WATTS: The point might also be made that a great deal of mystical literature is injunctive and is misunderstood by philosophers as being descriptive.

SPENCER-BROWN: Oh, yes. I would guess so.

WATTS: Take Patanjali.

Opinions and Knowledge

SPENCER-BROWN: Yes, people without the injunctive discipline in mathematics, apart from cookery and things that aren't generally admitted into the academic curriculum—mathematics is the only subject of any importance in the academic curriculum which uses injunctive language. And it is not chance that it is the only subject which doesn't deal in opinion. Because, in the use of injunction, it is not a matter of opinion what the result is going to be, you know.

And it's when we get very—these people who have been very sloppily educated; and they have, as we did back in England recently, a program on the television and they were all social scientists, and they said "Well, we have a measure of madness, and it is to know something." If you know it, you are mad, you see. If you only think it, well, that's sane. The great ignorance these people displayed, you see, is the ignorance of the queen of the sciences, as mathematics is often called. For example, let's take this book I was mentioning before, which is such a beautiful book in three volumes—Dickson's *History of the Theory of Numbers*. Oh, I don't know… there is 1500 to 2000 pages[9] absolutely crammed full—not a single opinion—it's all knowledge; it is all what is so. The gross ignorance expressed by these people, you see. This dealing in opinion can only be done by the ignoring of the disciplines of knowledge. Because, if it is an opinion, then it must be wrong—because if it were not so, if it were not wrong, then it would be knowledge, and it wouldn't be an opinion. So when—you know, when somebody comes and says "I think so"—well, that's an opinion. If you knew it, you wouldn't think it.

As for the other trick which is played, which is "You know, you don't know anything, you see, you don't know a thing, you know." You say, "Oh yes, I know what I had for breakfast. Oh no, you may have forgotten, you may have made a mistake." The proposition that such people produce is that anything—Russell, himself, was one of these. You know, he said "I don't even know that two and two make four. You see, I may have been mistaken ... ." It is put more cogently than I could put it, my heart isn't in it. He was laboring a point because it was necessary for his subsequent statement that he should establish this, you see. So the theory: "You don't know; you know nothing at all; it's all a matter of opinion." And, well, the question I always ask such people is "How do you know this? How do you know that nobody knows anything? How do you know it is only a matter of opinion?"

WATTS: Isn't that the same kind of a question, when you say to a relativist: "You mean that everything is absolutely relative?"

SPENCER-BROWN: Well, it is the same kind of throwing back his own system at him to show that he cannot support himself. There is the bland statement which really comes out in the form "I know that nobody can know anything."

MAN: That's the paradox.

SPENCER-BROWN: Well, all you have got to do is say "How do you know?"

MAN: I can't tell you.

SPENCER-BROWN: No.

O'REGAN: Some of your analysis of contradiction, and whole notion of crossing over from marked to the unmarked would almost suggest that contradiction, in a sense, is the form of form. It is what we can see when one arrives at that stage. Maybe the book could be the *Laws of Contradictions* just as much as the *Laws of Form*.

SPENCER-BROWN: Well, I am always careful about putting something greater into a smaller pot. You see, whenever we are speaking of contradiction, it is at such a more superficial level, because we are now already in language, and so on. Whereas in laws of form the form is operative at every level. Whereas contradiction is only operative in

---

9. 1601, in three volumes: 486,802,313

something like our speaking. That's why, you know, although it's illustrative, it wouldn't do as a substitute.

LILLY: In the act of creation, using a self-referential tunnel feedback, can you move from inward to outward on your five levels, or your orders here, or is this just restricted to the first distinction?

SPENCER-BROWN: I am not quite sure—you mean, "Can you distinguish the five eternal orders?"

LILLY: Right. One from the other, moving from one level to the other, using the self-referential feedback, in each case, so that you get an oscillation between the two levels.

SPENCER-BROWN: There's no feedback in heaven.

LILLY: OK. At what point do you create feedback?

Feedback at the Fifth Crossing

SPENCER-BROWN: When you go into the first temporal existence.

LILLY: So you have got to be on six?

SPENCER-BROWN: Five.

MAN: Going up.

SPENCER-BROWN: It's the fifth crossing.

LILLY: So the paradox does not appear until the fifth crossing.

SPENCER-BROWN: That's right. No, there is no time before that, and that is why they are eternal, the others. You think that it is going to. You don't know that it is going to happen, you see. You are coming out, you know—it's OK, it's still eternal, you know, you can still see the whole. And you get one too—You know, you get a little overconfident. Well, why stop here? Let's try going out a bit farther. Now where are we?

LILLY: You spoke of the fifth order equation as being runaway.

SPENCER-BROWN: That's in numerical mathematics, yes.

LILLY: Now, where do they apply here?

SPENCER-BROWN: Oh, they are not in here—this is just an analogy.

LILLY: This is outside them.

SPENCER-BROWN: Ya, the fifth order—You know, there is evidence all over the universe of a special state where you come to the fifth degree or the fifth whatever it may be, and bang, it changes.[10] It was all self-contained before that. This is a technical point mathematically in the question of solving equations for the varying degrees. You can solve degree one, this is ordinary numerical algebra. We can solve degree two. There is a formula, an algebraic formula, which most of us learn in school, for doing that. And by an extension of that we can solve degree four, also by an algebraic formula. I missed out three—we can do that, you see—and then the further extension of four. And everybody thought for quite a long time, I don't know just exactly when it was, not so long ago, that if only we could find this, we could find the formula for degree five equations—find the roots, and so on. In fact, we can't, because, without going into detail, something has been—something overtakes something else. Instead of your being able to reduce it to the equations of a lesser degree, you suddenly find that your reduction uses degrees that are higher degrees than you have already started with.

LILLY: It's an expanding system.

SPENCER-BROWN: It runs away, and there is no winning formula for finding the roots. You just have to find them ad hoc.

LILLY: So it becomes a partial feedback system.

VON MEIER: Divination.

SPENCER-BROWN: Ya, it runs away, it runs away. Before, you could get back by rule. After that, there are no rules for getting back. You may hit upon a rule, but there is no rule for finding it. It becomes more, you know, lots of rules of thumb. Like in the present existence—run away with itself a long way. And, you know, there are no formulae for getting back. There are a lot of ad hoc rules.

WATTS: We should pause to change the tape, James.

---

10. Humpty Dumpty had his great fall, from which he could not be put together again, on the chessboard's fifth crossing. *Through the Looking-Glass* is instructive in following the discussion the levels of the cosmos. Alice starts on the zero square, and takes her first step through the mirror (Monkey dives through waterfall) landing on Q2—the second square, cardinal number one, in the marked state behind the glass.

SPENCER-BROWN: Well, is it finished? Nice meeting you. We have been spinning it out, but in the end there is really nothing to say.

LILLY: Well, thank you very much for coming all this way to talk to us, and we hope that you will come back.

WATTS: If you'll come back for a more leisurely session.

SPENCER-BROWN: Well, I hope it can be yes.

MAN: The beginning of the end.

LILLY: We've gone from zero to—

End of tape.
End of Session Four.

Snelson, P., II. (2015). *Cybernetic Square Dance*. Computer graphic.

Snelson, P., II. (2018). *Notre Dame on Fire*. Computer graphic.

Commentary: The First Message From Space

Kurt von Meier and Cliff Barney[1]

Lave a whale a while in a whillbarrow (isn't it the truath I'm tallin ye?) to have fins and flippers that shimmy and shake. Tim Timmycan timped hir, tampting Tam. Fleppety! Flippety! Fleapow!

Hop!

—James Joyce, *Finnegan's Wake*

I. The Wanderings of *Man*

This commentary on the transcript of G. Spencer Brown's remarks at Esalen Institute March 19-20, 1973, draws inspiration from Karl Pribram's suggestion that it be cast in the form of a Sufi story, which teaches by its own example. Upon examining the text, we find that, for those who care to cross into the vision of the morality play, the voices of Lilly, Bateson, Watts, Von Meier, and the others, particularly the universal, anonymous questings of Man, ever seeking to get it just right, make up, with the patient, measured, precise explanations of Spencer-Brown, the touring champ, the antiphonal chorus accompanying the unveiling of the mystery. We experience G. Spencer-Brown's informal discussion of the laws of form as one of the possible manifestations of the long-awaited first message from space. It is about Time.

We call attention to Brown's discussion (Session Two, Prime Numbers) of the properties of the number 2311, prime factorial plus one for 11, which is the fifth prime and the number whose powers give the coefficients, or constant aspects, of the binomial expansion, $(a + b)^n$. Melvin Fine, the Hasidic mandarin, has reminded us that Leibniz knew these same coefficients to govern the hexagrams of the I Ching, according to the number of yin and yang lines in each group. It did not cross our path until much later that 2311 is also the index to the sequence listed by N. J. A. Sloane (1973), in the invaluable *A Handbook of Integer Sequences*[2] as a friendly beginning for communication with Betelgeuse. Sloane calls the name of sequence 2311 as *non-cyclic simple groups*. Brown refers to it as *Big M Plus One*.[3]

---

1. Email: cbarney@jeffnet.org
2. There is currently a second hardback edition and paperback edition, plus an online eTextbook.
3. We note here the remarkable coincidence that Big M Plus One, or 2311, is reflected as 1132 A. D., when "Men like ants or emmets wondern upon a groot hwide Whallfisk which lay in a runnel," on p. 13 of *Finnegans Wake*, by Joyce, James (Viking Press, New York, 1955). We have elsewhere discussed the reflection of the other Joycean year, A. D. 566, as 665 A. D., just after the Synod of Whitby, at which King Oswy of Northumberland bowed to the wishes of his Kentish queen and her Roman advisors and accepted the solar reckoning for the date of Easter; thus cutting the cord, as Yeats put it, that bound Christianity to the Druids, and setting the course of the Western yang empire.

Betelgeuse being in the armpit of Orion, we might look for a glyph of this message in the figure of the first martyr, according to the Western tradition St. Puce, the flea reported by hearsay to have been impaled by the Roman centurion's merciful spear. However the tradition is perhaps somewhat richer if we focus on Orion as the (blind) Wanderer, which—writes star cataloguer Richard Hinckley Allen[4]—we may do. Here, as Dr. Fine reminds us, we meet Lü, the Wanderer, hexagram #56 in the subtle ordering of the I Ching offered by the Duke of Chou, or #13, the number of Dionysus, in the simpler binary notation of Boolean logic.[5]

The Wanderer is also Wotan, Odin,[6] through whose 540 doors went 800 warriors each to fight the Wolf, thus making up a force of 432,000 warriors: one for each year, we learn from Joseph Campbell, of the reign, between the Creation and the Flood, of 10 mythical Sumerian kings, and one-tenth the number of years in the Hindu kapla. Campbell (1970) argues persuasively that these numbers refer to the precession of the equinoxes. The Sumerian number system was based on the *soss* (60, the second term in the series as published by Sloane, though actually the first in the sequence of non-cyclic simple groups), and 432 x 60 = 25,920, the number of years required for the earth to make one complete nutational wobble, one trip through the Zodiac, if we figure the precession at the Sumerian rate of 50 seconds of arc per year, or 72 years per degree.[7]

Campbell also points out that in the Hebrew tradition the years of the patriarchs from Adam through Noah (until the Flood) total 1,656. We may hypothesize that James Keys, the yin persona of the Brown guru, had this in mind when he published *23 Degrees of Paradise* (1970).

The number 54(0) we associate with Shakuhachi Unzen, who practices the 54 steps of his yang form T'ai Chi Chuan exercise as he mops the floor and removes the garbage from the Teahouse of Necessity, where every night is served an installment of the Feast of 4001 Fools. His numerical token is actually 54.54, based upon a transformation of the standard unit of measure employed in ancient Japan, in

---

4. Orion as Wanderer is revealed in the constellation's earlier name, $Αλητοποδιοω$, from $Αλη$, roaming, according to Allen on p. 304 of *Star Names/Their Lore and Meaning* (a 1963 Dover reprint of *Star-Names and Their Meanings*, first published by G. E. Stechart in 1899). The name *Orion* is from $Ωαριων$, which is cognate with English *warrior*.

5. Hexagram 56: Lü, the Wanderer, is shown below

═══ ═══	above: Li, The Clinging Fire	The six broken and solid lines of the hexagram, read from the bottom, nay be mapped into binary numbers as 001101, or 13 to a decimal base.
═ ═ ═ ═	below: Ken, Keeping Still Mountain	

Mathews translates the Chinese character for Lü (#4286) as "A guest, a stranger. To travel, to lodge."

6. Wotan appears as the Wanderer in *Siegfried*, the third opera of the Wagner's Ring cycle. Siegfried brushes the Wanderer aside on his way to awaken Brünnhilde and do something about the Rheingold.

7. Modern observers put it at 50.27 seconds per year, for a 25,780.783 year round trip

particular the tradition that 1.8 *shakus* determines the standard tuned length of a bamboo flute (made, perhaps, of bamboo from the very grove in which the teahouse is situated). Such a flute might be measured at 54.54 centimeters, by those following the units established by Napoleon Bonaparte after his mystical visit to the Great Pyramid of Cheops, having entered the King's chamber by torchlight, alone and at midnight (in meditation, we may speculate, upon the nature and action of Amoghasiddhi Buddha: black hat, green eyes, crossed double rDorjes). Napoleon never did talk about what he experienced until late in his life in exile (latin *ex(s)ul*, wanderer) on St. Helena, wasn't it, the island named for the mother of Constantine (arithmetic being about constants, as Brown explains in the text, Session Two). Then he declined to tell the story, saying "You wouldn't believe it anyway."

What story? What sort of message can we read in this rendering of the spoken word of the Brown messenger? We note that the oracle of the tortoise is specifically referred to on several occasions (Session Two). During the Shang Yin dynasty (1523-1027 B. C.) the priests of the royal court divined secrets by heating the shell until it cracked and then noting how the cracks intersected the 13 interior and 25 circumferential divisions grown naturally by the tortoise. When the written language went public, the oral tradition went into silence, exile, cunning. For public consumption we are referred to the vegetable oracle, the yarrow stalks of the I Ching.

The tortoise gives us the transformation to a new constellation, Lyra, known as the Little Tortoise, or Shell, "thus going back to the legendary origin of the instrument from the empty covering of the creature cast upon the shore with the dried tendons stretched across it" (Allen, 1963, p. 283). This is later Apollo's lyre, seven-stringed, discovered by Hermes, inventor of dice, three of which can permute the 56 minor arcana (Wotan again) and two the 21 major arcana, which, with the Fool, who is *hors commerce*, make up the deck of cards known as the Tarot, whose name has the same root as *tortoise* (Indo-European *wendh*, to turn, wind, weave), which is also the root of *wander*. One can pick up the skein at any point and begin to tease out a thread that will lead one through the labyrinth.

"This is indeed amazing." So wrote Brown (1969, p. 105) upon surmising that the world we know is constructed "in order (and thus in such a way as to be able) to see itself." We are referred to the maze at Knossos, where the spider lady, Ariadne, had Theseus dangling by a string. We take hold of the string: Labyrinthos, from *labrus*, not Greek but a Minoan word for double axe. And we use the iconic representation of the double axe as a framework upon which to construct the map of the labyrinth (see Richardson, 1966, pp 285–296, esp. p. 291). We are being told here that all representations of life, for example, stories, myths, fables, pictures, are explicitly representations of themselves, self-referential. Re-presentations. And the thread we hold, that leads us out of the labyrinth, where no path seems to go anywhere, since they all come back on themselves, is language; we follow it to learn the order of the labyrinth. With the text (Indo European *teks*, to weave, fabricate) as a clue to the identity of the Vajrayogini as the mythical Joyce James, one can anywhere and everywhere pick up the thread upon which Brown has strung the pearls he has divined.

II. Points of View

Nevertheless, to make an expression meaningful, "we must add to it an indicator to present a place from which the observer is invited to regard it" (Brown, 1972, p. 103). These pages reflect in one facet of the crystal mirror the vision of the real (meaningless), the true, the false, and the imaginary, mapped out on the plain crossed by G. Spencer-Brown. The text, for instance, purports to be a transcript of the words Spencer Brown spoke at the South Coast Motel, a mile or so down the road from the sulphur baths at Big Sur, California, on the occasion of the AUM conference conceived by Alan Watts and John Lilly (the master passing over the pole on the occasion of the last full moon before the vernal equinox). Confirmatory evidence for this viewpoint, definitely northern hemisphere (London) may be obtained from tapes of the conference (should anyone be able to find them, nearly a half-century after the occasion). Lucky listeners to these tapes can verify that Spencer-Brown's remarks are reproduced here with close, if not perfect, fidelity.

Arriving at Cape Town, however, we cast the net tied by N. J. A. Sloane and catch the icon of an all-of-a-piece, multidimensional message to inner space—that is, the space of the eternal regions, where we find numbers, including the number represented, in the Arabic notation, as 2311. Here nothing is hidden, since the space is created "until the time came for time to begin," as Brown (Session One transcript, this issue, p. 30) says C. E. Rolt says in his (Rolt's) introduction to the *Divine Names* of Dionysius the Areopagite, and the eyes of the (one-way) blind are opened and the ears of the deaf unstopped. Messages from and to space are all around us, and we have only to read them. "In nature are signatures/needing no verbal tradition,/oak leaf never plane leaf" (Pound, 1970, p. 573). Caught in the net of number, Big M Plus One provides name, rank, and serial, as required by convention, and stands revealed in an entirely different role, the index of non-cyclic simple groups and thus a nodal transfer point to the one-eyed Wotan, who braided the hairs of the Night Mare's tail.[8] We tie the net one knot at a time, and from any node, 2311 leads us to infinitely numerable license plates, phone numbers, grocery bills ($23.11) and the like.

Wotan (Woden's Tag/Wednesday/ midweek/balance) carves the runes on a staff cut from Yggdrasil, the world ash, which binds together earth, heaven, and hell, branches mirroring roots, which we may allude to, but may not uncover without killing the tree. Sloane (1973, p. 12), points out the mathematical aspect of trees, rooted and otherwise, as graphs containing "not closed paths" (unlike the net, in which we catch the icon). For the English language, a knowledge of which is assumed by Spencer Brown on the part of a reader of *Laws of Form*, the roots are named with psychedelic clarity in the appendices to *The American Heritage Dictionary*.[9] In these pages, the (conjectured) Indo-European and, lately, Semitic sources of the language are cross-indexed to the common words of the vocabulary; so that, for instance,

---

8. See "The Theory of Braids," by Emil Artin in *The American Scientist*, Vol. 38, No. 1, pp. 112-119, January, 1950, and other references cited for the article "Group Theory and Braids" in Martin Gardner's *New Mathematical Diversions from Scientific American*, Simon & Schuster, New York, 1966.
9. New York: Houghton Mifflin.

having been referred from *tree* to the Indo-European entry *deru-*, meaning firm, or solid, we find the collapsed meaning of *tree, truth, Druid, trust, trough, troth, durable,* and so forth. The path leads through Greek *drus* (the d-t shift having been found out by the Brothers Grimm, who knew that fairy tales were about language, self-referential): oak for Oakville, Napa County, wine valley of Dionysus, where, at 7700 St. Helena Highway (named for the mother of Constantine, wasn't it?) under 700-year-old oaks this mad commentary is being written.

Turning scales to feathers, like dinosaurs, we take flight for Christchurch, entering the imaginary state in which Shakuhachi Unzen, Woody Nicholson, and Primo the Fool braid their destinies, and in which these pages are a program note for the great Chaco Canyon Eisteddfod of 1976.

This sitting of poets, musicians and shamans provides a mythic counterbalance to the electoral process, which, as the United States marked its 200th anniversary having lost, or forgotten, its commitment to due process, showed signs of breaking down (in accordance with the instructions formally built into the Constitution by the 55 Freemasons who wrote it one summer in Philadelphia, on the 40th parallel). Unzen it is whose Sufi listening post in the heart of California (state named for the white goddess of Don Quixote's impossible dream, she to whom all poetry is addressed, according to Alun Lewis, according to Robert Graves, attest G. Spencer Brown as James Keys (1972, p. 107–108), is the scene of the appearance of the Yellow Pearl, who, as the 12-year-old T'ai Chi star of the touring Peking Opera, becomes an international cause célèbre and focal point of the Eisteddfod.

Shakuhachi's guide and guru is the busboy who, in the same teahouse for which the monk serves as Cold Mountain Pratyeka janitor, each night concocts a special brew which bears the name Ti-Tseng. This he mixes after all the ordinary guests have been served, and the choice can be made of the person who shall reign as the King (or Queen) of the Fools aboard the space/time, gravity/gracewarpship Adamantina, into which the teahouse periodically transforms all within its Mandala of energy/void. The stuff in the pot of Ti-tseng turns out in the clear light of the Bright Early Morning Star to be dregs collected from all the partially emptied receptacles returned on his tray to the scullery. To this much the regulars and old-timers are hip; indeed, the service of Ti-tseng is ritualized, as when the jaguars, bursting into the temple and desecrating the altar, repeatedly, through generations of the priesthood, become elements of epiphany in the Kafkesque Tantra. Now there is the bank of watchers, who ever decline the sip, but appear faithfully at the subsequent convocation: latecomers, fruhsplitters, Yzakers & your 1 in 10 schizoid, prepared for the culmination of yet another one of the fantastic (but finite) Feast of 4001 Fools (dishes prepared for the Uwaysiyya Sutis, 4001 of whom are said to be wandering the face of the earth at all times, without credentials from Soofi Central, the Bodhidharma Certification Board, or the Magister Ludi contest committee.)

The Eisteddfod is arranged by the country's No. 22 vice president, Woody, that he may receive advice on how best to proceed, he having been thrust into a position resembling power when the president and the first 21 vices were wiped out at Ahab

McGaff's Double Cross Saloon in Las Vegas (Vega being the alpha star in Lyra). Woody, who travels from national park to supermarket parking lot in the last of the Winnebagoes, a superbly outfitted camper equipped with heliport, swimming pool, and satellite reception station, is assisted by the mysterious agents of the Sufia, among whom we note the Bodhisufi Bismullah Tariq and Melvin Fine, the Hasidic mandarin.

In this context Joyce James appears as an aide to the llama Al Paca, liaison man between the Sufia and the spiritual materialist arm, MaFie, or MyFee, which is simultaneously trying to fix the eistedifodd in favor of Ahab and his all-girl security force, the Kritiquettes, and, through Ahab's Miasma Beach cousin Jackie, to produce it for global television. J. James discovers the Spencer-Brown text while researching the form of distinction that appears in the medieval vision of quests, which were tragedies, and pilgrimages, which were comedies (see Holloway, 1974). The document presents the Joycean hypothesis that *Laws of Form*, with its demonstration of the generation of Time, offers a means of mapping cultural transformations which themselves reflect our own transformations as refugees in Time.

Brown's performance at Esalen certainly earns him a place in the Eisteddfod finals, along with Ahab's black vision, the entry from the teenage author of essay #768 in the Working With Negativity Sweepstakes, the taped hoax perpetrated by members of the Imaginary Liberation Front, and the dance of the Yellow Pearl herself, our first female leader—who, being uncommitted to any particular truth, replaces government by control of information, secrecy, and deceit with leadership through inspiration, education, and enlightenment, calling to account our kings, corporations, imaginary persons before, and frequently above, the law.

Meanwhile Primo puzzles out the controls of the Adamantina, which have been locked on a course for the Black Hole in Cygnus by the captain, Jetsun Rainbowshay, who has vanished. In the crystal navigation table, Primo finds maps to the cosmos—*Laws of Form*, the tortoise oracle, the I Ching, the Tarot, the dice of Hermes—which he can read only with the help of Melvin Fine.

III. The Adamantine Isomorph

These three stories fit together as the net, tree, and array structures for which the present text provides the transformation rules. The terms were introduced to us by Christopher Wells, Center for Music Experiment, University of California, San Diego, in an unpublished manuscript dated 9/22/74:

- Web [net] grammars map (iconic descriptions) onto directed graphs [digraphs, i.e., "trees"] or (iconic descriptions).
- Tree grammars map (iconic descriptions) onto hierarchical directed graphs or (symbolic descriptions).
- Array grammars map matrices of point terminals onto transformed...matrices or (images). And the key:
- Metagrammars map the other three grammars' elements onto each other as plan/command learning or (extensions).

The adamantine isomorph! To its four elements we may conventionally assign air/earth/fire/water, Matthew/Mark/Luke/John, Prometheus/Epimetheus/Atlas/Menoetius (the four Titan brothers who warred on Cronos with varying success), and so forth. We know from the laws of form that once we make any distinction whatever we generate inevitably the eternal archetypes. The forms are always the same and every story is the same story, which is about Time. History, from Greek *histor*, wise, learned, from Indo-European *wid-tor-*, from *weid-* to see. The image and its mirror; sun and moon.

IV. Return to the Form

The Father, Son, and Holy Ghost sitting around Mary's place one day, weighing contemplations upon Time, which is the inevitable harmonic of all that moves, whirls its way through the mother space; and her body quivered with the energy, as born from the star; and the message transmitted through the body, the soma, into the form of the kingdom of heaven, the way seen through Pearl/Gate with one eye, bloodshot, close-to for about an hour in Buenos Aires (the piece by Marcel Duchamp in Noo Yawk's MOMA). *Om gate gate paragate parasamgate bodhi svaha.*

She say, "I want to get in lights, front billing instead of the shadow me, nobody. Bear me, as a star." And the swastika in the sky began to spin...Santa Lucia on the South Sur coast; in the Sky with Adamantinoids, cast to the serpent in the peaceful ocean to appease the Wrath of War on the surface, beneath the peaceful brow, the lambda curl of the wave, shaken from within the earth, in one of her periodicities, the signature for which, her handwriting, is the sequence indexed 2311 by N. J. A. Sloane of Bell Labs. Lucy in the Sky, with her pet Beetle Goose. The sign in the sky for Constantine, the Paramahamsa, the Great Himalayan Goose or Gander! Goosey-goosey Gander, where shall you Wander?

They say, "All right, MS! [Mother Superior] We give you the cultus, see? We champion Mariology. We do everything but spell it out."

Mary: So start spelling out the sign & beans.
HG: We're sniffing, Apollo-like, lupine around da point here and coiling upon an oomphalosity white tower of ivory, such as that coveted by the hungry ghost of Ahab, who, we three Kings remember, ruddy-coiled out into the well drainage at the Good Lady (was it?) Samaritane (SOLDS!) ... in the bottom of the chariot, cancer, whirlpool, in the depth of the Form. And so we fain preserve the whale. There shall be no personal rejection of the life of great beings. No more whales, elephants, dolphins to die. That is the tip of the balance, the stated, stipulated bias toward compassion, which places others above ourselves, dedicating the benefit, if there be any, of this meditation to others. To the kings of the air, the generation of brave eagles who hunt the jet planes, kamikaze! Bees, humming birds, the red tail hawk, PALAKWAIO, Simurgh, Garuda bearing the Buddha of all the Buddhas, carrying the standard, from out of the dismal maze of the men in their fury contending for the right!
Father: There. We now have the Holy Spirit out into the Marked State. Memory and genetics, the arithms as perceived by our senses and programmed into our biocomputers. What seem to be invariants: K is for konstant. Now there's no sense

everybody getting out at once. If Mary wants out into the Marked State, then who goes back into the Form to keep it all balanced?
Son: Me again?
HG: Look, I been in there long enough. Anyway, wasn't the logic of it me coming out, so that all secrets should be revealed?
Father: Hmmmm. What would Mr. Natural do in such a situation?
Mark: Don't chicken out, Baba, golden fleece.
Mary: That's a good one of my boys.
Father: Looks like it's my turn. *Om doubt.*

References

Allen, R. H. (1963). *Star names, their lore and meaning.* New York: Dover. (Originally published in 1899)
Campbell, J. (1970). *The masks of God: Oriental mythology.* New York: The Viking Press. (pp. 115–130).
Holloway, J. (1974). *Figure of pilgrim in Medieval poetry.* Unpublished doctoral thesis, University of California, Berkeley.
Keys, J. (1970). *23 degrees of paradise.* Cambridge, UK: Cat Books.
Keys, J. (1972). *Only two can play this game.* New York: Julian Press.
Pound, E. (1970). Canto LXXXVII. In *The cantos of Ezra Pound.* New York: New Directions.
Sloane, N. J. A. (1973). *A handbook of integer sequences.* Cambridge, MA: Academic Press.
Richardson, L. J. D. (1966). The labyrinth. In L. R. Palmer & J. Chadwick (Eds.), *Proceedings of the Cambridge Colloquium on Mycenaean studies* (pp. 285–296). Cambridge: Cambridge University Press.
Spencer-Brown, G. (1969). *Laws of form.* New York: Julian Press

Snelson, P., II. (2010). *Field of Matrices in White.* Computer graphic.

Snelson, P., II. (2015). *Dancing Inside the 2-Dimensional Picture Plane*. Computer graphic.

Snelson, P., II. (2012). *Structured Scribbling Maniacal Matrix Graphic Gestures*. Computer graphic.

The Flagg Resolution Revisited

James M. Flagg[1] and Louis H. Kauffman[2]

This paper discusses the Flagg Resolution for resolving paradox: If a logical element is equal to its own negation, as in $J = \overline{J}$, then in any given formula, any substitution of $\overline{J}$ for J must be made for either no occurrences of J in the formula, or for all occurrences of J in the formula. The paper explains the background, meaning and applications of this method for paradox removal.

I. Introduction

An early virtual logic column (Kauffman, 1999) in *Cybernetics and Human Knowing* was devoted to the Flagg Resolution. This paper revisits the subject. The Flagg Resolution (FR) is a way to handle a paradoxical entity when it arises in mathematical language. Put in a nutshell, the FR avoids contradiction by modifying substitution to save the truth.

In the cybernetics context we are familiar with handling circularity in systems of thought and praxis. Formalisms for circularity, such as a self-pointing arrow or a form of reentry as in the re-entering mark illustrated later in this paper, are often taken to exemplify the cybernetic way of thinking. This way of thinking includes the observer/participant in the system and does not make the standard distinctions between map and territory, between formalism and what is formalized.

On the other hand, when we work directly with a self contradiction such as the paradoxical statement "This statement is false," we are aware that the usual handling of such statements by classical logic leads to confusion. Is the self-denying statement meaningful? Can it be regarded as true or false? George Spencer-Brown, in his book *Laws of Form* (Spencer-Brown, 1969), suggested that the way out of confusion about such statements is to understand that they are happening in *time*. At a given time the statement is True, but it says that it is False, so it must be False and that is its value at the next time. But now it is False and it says that it is False, so its value at the next successive time is True. The statement undergoes an oscillation of value in time:

... *True, False, True, False, True,* ...

And this oscillation does not cause us to change our logic, only to apprehend the temporality inherent in such apparent paradox.

Once such temporality is recognized we can see that logical relations remain as before. For example, let S be the statement "This statement is false." and let $\sim S$ be its

---

1. Email: jamesm.flagg@gmail.com
2. Louis H Kauffman, Mathematics Department, University of Illinois at Chicago, 851 South Morgan Street, Chicago, Illinois 60607-7045. Email: Kauffman@uic.edu

negation. If S oscillates as above, then $\sim S$ oscillates oppositely and we can examine them together as in $S \vee \sim S$ where $\vee$ denotes "or."

S: ... True, False, True, False, True,...
$\sim S$: ...False, True, False, True, False,...

Then we see that at any given time

$$S \vee \sim S = True \vee False = True$$

or

$$S \vee \sim S = False \vee True = True$$

and thus we can conclude that $S \vee \sim S$ has the value True in any case.

In fact, we encounter paradoxicality whenever we push the limits of language and observation. For example, it is common to think that the cup on my table is a localized object that can act as a container. Little reflection is required to realize that the cup is in fact determined by everything that it is not! And what the cup is not is the rest of the entire universe. Any thing is identical with what it is not.

Figure and *ground* determine one another.

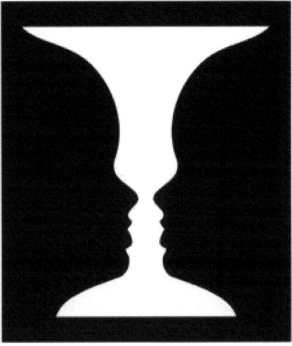

Drawing of Vase and Faces.

We apprehend, in the well-known illustration of *vase and faces* that *faces and vase co-determine* one another. And in speaking of the form of the *drawing* we understand that the *drawing* is both *vase and faces*.

The *drawing* is what it is, but *vase and faces* are the two *faces* of that *drawing*: Vase *is* faces; faces *are* vase. A thing is what it is not.

Speaking cybernetically and experientially we are not confused by such speech. But the logical speech of separate distinctions seems to demand something more. The logical speech does not seem to want to allow us to say that the *vase* is identical with what it is not, nor that the *faces* are equivalent with what they are not. The logical speech says that a thing and what it is not fill the logical space and form the Truth.

The purpose of this paper is to present one view and its ramifications about how to heal this split between cybernetic/experiential thinking and logical speech.

Given that we denote negation by $\sim$ so that "not X" is "$\sim X$," the quintessential paradoxical entity is an entity J that is identical with its negation, identical with what it is not. We write $J = \sim J$. Such J must be considered when thinking widely and when thinking of the world and most particularly in thinking of fellow human beings who are indeed the soul of contradiction. They are what they are not.

It is the relationship between the thing and what it is not that makes the Whole and the Truth. In *logic* we say that for any statement P, P together with $\sim P$ gives the whole universe, gives the truth and we write

$$P \vee \sim P = \text{True}. \text{ We use } A \vee B \text{ to denote ``}A \text{ or } B\text{."}$$

Let us take the logical point of view. When we write $A = B$ we intend that A and B are in some way identical. Thus we can write *vase = the white region* and we can write *faces = the black region*. And with this specificity about things being things to themselves we imagine that these things can be replaced, if we wish, by their equivalents. Thus in ordinary mathematics I may write $2/3 = 4/6$ and therefore

$$2/3 + 1/6 = 4/6 + 1/6 = 5/6.$$

Such are the acts of mathematical work.

Keeping to this notion of the ubiquity of substitution, I can see how the equation for the paradoxical element $J = \sim J$ will fare as a currency for substitution. Consider the statement

$$\text{True} = J \vee \sim J.$$

We have agreed upon this Truth on the general ground of logic.

Remember that we have invited the devilish J to participate in a substitution. We shall substitute J for $\sim J$ and obtain

$$\text{True} = J \vee \sim J = J \vee J = J.$$

Thus True = J. But then False = ~True = ~J = J. So we have False = J. Since True = J and False = J, we conclude that True = False!! This is the collapse of *logic*. What is to be done?

Here is an imaginary conversation with James Flagg on the matter. Flagg is real. This conversation is imaginary.

LK: It seems that I cannot have an entity such that J = ~J without collapsing logic and creating serious inconsistency in the way we speak.

Flagg: Attend to it more carefully. Right now J is whatever J is. When you write $J \vee {\sim}J$ then at this moment if J is True then ~J is False, and if J is False, then ~J is True. So there is no contradiction. $J \vee {\sim}J$ is indeed always True. It cannot suddenly become False.

LK: So you are telling me that I cannot make an arbitrary substitution of J for ~J?

Flagg: Certainly not! I'll tell you what you can do. If you have an expression like $J \vee {\sim}J$, then you can make the substitution for both instances of J in the formula. This preserves the relationship between them.

LK: So I can write $J \vee {\sim}J = {\sim}J \vee {\sim}{\sim}J$. I see that this is innocuous. After all, ${\sim}J \vee {\sim}{\sim}J = {\sim}J \vee J$, since ${\sim}{\sim}J = J$ and I am sure you will allow this substitution.

Flagg: That is correct. We have a paradoxical element J = ~J and the key substitution of ~J for J throughout, represents its fixity, or fixedness, if you will, under negation, truth, and indication (which here are a portmanteau ensemble). This substitution must be performed for all appearances of J throughout a formula, *salva veritate*, as Leibniz (1677/1998) would have it, or not at all (see Leibniz, 1966 for a discussion of substitution). This dictum is the Flagg Resolution. I will say more about this in a separate section of our paper.

The imaginary conversation between LK and Flagg will continue in the next section of the paper. We interrupt it here to indicate how the paper is organized. The sections of the paper are:

I. Introduction
II. Continuing the Conversation
III. The Formalism of Laws of Form
IV. James Flagg on the Flagg Resolution
V. Einstein/Penrose Tensor Networks and the FR.
VI. The Russell Paradox.
VII. Epilogue

The conversation in the Introduction continues in section II. Section III provides more background on *laws of form* and its interpretation for logic. In section III two new participants join the conversation. These are Cookie and Parabel, who are sentient text strings precariously existing very near the void. They have appeared in other writings of LK (e.g., Kauffman, 2017) and add to the discussion by their very grounded insight into matters epistemological. Section IV is an insightful essay by James Flagg on the Flagg Resolution. Sections V and VI continue the discussions with Cookie and Parabel on analogs of the FR in tensor networks and knot theory, and in section VI we discuss the Russell paradox both from the point of view of the FR and from the point of view of Goedel, Bernays, von Neumann set theory (Mendelson 1997), where a *type* distinction between *sets* and *classes* banishes the paradox. We compare that with the FR and its way of living with the paradox as a form of being. Section VII returns to the main theme of truth and substitution. By allowing the Flagg Resolution we can stand under logical *circularity* and begin to understand the cybernetics of reason and the reason of cybernetics.

We dedicate this paper to George Spencer-Brown, Francisco Varela, David Van Cleave Lincicome, and John Ewell. And we wish to thank Providence for the many opportunities that have been provided for discussion and reflection on the issues brought forth in this paper.

II. Continuing the Conversation

LK: How do we know when an element is paradoxical? For example, I can have the famous Russell Set

$$R = \{x \mid x \text{ is a set that is not a member of itself.}\}.$$

We discover a fixed point for negation in the form of the question: Is R a member of itself? According to the definition of R, R is a member of itself if and only if it is not a member of itself. The statement J = "R is a member of itself" has the property that if J is True, then J is False, and if J is False, then J is True. So we can apply the FR to J!

Flagg: And you are worried that perhaps R is not a well-formed set since it seems to flicker with respect to its own membership in itself.

LK: Yes. Perhaps we have to change other aspects of the discourse about sets in order to be sure there are no further contradictions.

Flagg: Ah! You have not got the whole picture. I do not attempt to create mathematics that is free of all contradiction. If there are contradictions there will occur instances of J with $J = \sim J$. Whenever this happens, the J is treated specially with regard to its fixedness under negation. That is all we do. Otherwise life and logic go on as before. There are no new special logical values, and nothing is forbidden to the discussion.

LK: I want to analyze systems and formal systems to see if they are consistent. A consistent system cannot produce a J with $J = {\sim}J$.

Flagg: After the discovery of FR, whose discovery emerged from correcting a misconception of the great Francisco Varela in his "Calculus For Self Reference" regarding Spencer-Brown's

$$F = {\sim}F,$$

I never went back and revisited a single paradox, as I knew its resolute power. I knew it wiped out 2800 years of paradox. FR is what the ancient Egyptian and Vedic gods do. The only difference is that we now have an actual notation for their skullduggery. The ancient gods are living functions, or mythic functions, who fulfill all the requirements of Spencer-Brown's axiom of boundaries: A thing is what it is not. Even Wittgenstein said as much: "For instance, when I say: such and such a point in the visual field is blue, I know not only that, but also that the point is not green, not read, not yellow and so on. I have laid the whole colour-scale alongside simultaneously" (quoted in Waismann & McGuinness, 1967, pp. 63–64)

And so, the Vedic gods, or mythic functions, sit at the boundary (Being) between Asat (Non-Existence) and Sat (Existence). For ancient Egypt, Zep Teti (The First Time) took place inside the everlasting living waters of Nu as Atum slumbered. He, in an act of supreme self-reference calls his own name and moves from one state to another. In Being, Sat and Asat lay side by side, Nu and Atum move together. It's just as in Jimi Hendrix: "Good and Evil lay side by side, while electric love penetrates the sky." It's not new, but old: implicate order, wholeness, Tao. Or, according to De Nicolas (1978) the Vedic Sages say in "The Creation Hymn" in the *RG Veda*:

> That One which had been covered by the void ...
> The Sages searching ...
> Found in non-existence the kin of existence ...
> Who therefore knows from where it did arise ...
> He who watches in the highest heaven,
> He alone knows, unless ...
> He does not know. (Nicholas, 1976, p. 229, *RG Veda* 10.129)

I have provided here an elision of their poetic visions in which they bound duality and complements together. Once anyone thinks about God, or Love, they exist. So, we now understand why the negative postulation "Don't think about elephants" doesn't work. Any negative postulation posits its counterpart and complement. This is the source of all paradox.

LK: So in FR, *both* is neither a superposition nor a contradiction, it just is.

Flagg: Yes. It is common enough in everyday life. The most interesting entities are round squares. They should not be banished. And what do you do when you meet them in mathematics. You declare new worlds! The square root of negative unity is both +1 and -1 and eventually mathematicians learned to use it by endowing it with the symbol $i = \sqrt{-1}$, and explaining that $ii = -1$ and that i stood at right angles to the line of real numbers. The square root of minus one crossed the boundaries among number, algebra and geometry, and became a rotation of ninety degrees. This is a specific action of FR. We shall see more of this.

LK: Consider the sentence S that says "If this sentence is True, then unicorns can fly." We see that if S is False then it would be of the form "False implies unicorns can fly." and since an implication of the form "False implies P" is necessarily True, it would follow that if S is False then S is True. So S cannot be False. Therefore S is True and hence unicorns can fly.

Flagg: That is an admirable proof that unicorns can fly.

LK: But I could have also used the same form of proof to show that unicorns cannot fly!

Flagg: Indeed you could. Does this worry you?

LK: Of course it does. I can use this method to prove any statement that I like. It renders reasoning inconsistent at all levels.

Flagg: Of course you are making a mistake. "A implies B" = $\sim$A $\vee$ B as we all know. (When A is true then B follows.) So your statement S is of the form S = $\sim$S $\vee$ U where U is the Unicorn Statement. You are trying to avoid S = $\sim$S but you cannot do that without being inconsistent. It was by assuming that S and $\sim$S are distinct that you found yourself concluding the Unicorn Statement. You don't want that and so you must accept that S = $\sim$S. Then there is no problem. Indeed if S is False then S is True, and if S is True, then S is False. Just apply FR in using S.

LK: I shall have to think about this.

III. The Formalism of Laws of Form

In thinking further about the Flagg resolution it is useful to have the formalism of *Laws of Form* by G. Spencer-Brown (1969). In this formalism we have a mark, $\rceil$, that represents the distinction that this mark makes between its inside and its outside. We may think of the mark as referent to a single distinction that is given and sometimes called the first distinction. The mark is also seen as a transformation from the state indicated on its inside to the state indicated on its outside. Thus the empty mark we have drawn above can be seen as a transformation from the unmarked state on its inside to the marked state on its outside. We can further notate this by $\overline{u|} = m$ where u denotes the interior unmarked state and m denotes the exterior marked state. By keeping the language as parsimonious as possible, we allow u to be replaced by nothing since u denotes the unmarked state. Thus we can rewrite $\overline{u|} = m$ as $\overline{|} = m$, and so we see that the symbol for the marked state can indeed be taken to be the mark itself. We agree that "The value of a name called again is the value of the name" (Spencer Brown's law of calling), and so we can write mm = m and uu = u since each of these constitutes a repetition of a name. Then mm = m becomes ⁊⁊ = ⁊ and this equation is the version of the law of calling that is expressed by the mark. Each mark can be seen as the name of the other mark. Each mark can be seen as the state resulting from crossing from the unmarked state.

If we iterate the transformation we see that crossing from the marked state results in the unmarked state: $\overline{m|} = u$. This follows from the fact that we are describing the two and only two states related to a first distinction. One can cross from the unmarked state to the marked state and crossing from the marked state yields the unmarked state. What is not marked is unmarked. Now, unpacking the equation above and using the mark for m and nothing for u, we find ⁊| =. This is the law of crossing—"The value of a crossing made again is not the value of the crossing"—written in terms of the mark. In this way, consideration of one distinction leads to a calculus of indications, a language using only the one mark and this language has the two basic rules:

$$\overline{\overline{}} = \overline{},$$

$$\overline{\overline{}} = .$$

More complex expressions in the mark reduce uniquely to either the marked state or the unmarked state. For example, ⁊⁊⁊ = ⁊⁊ = ⁊. The reader may wonder, what does it mean to have two expressions E and F, standing next to one another as in EF or ⁊⁊? In this calculus of indications each expression stands either for the marked state or the unmarked state. We have that mm = m and uu = u. What about um?

Since u can be replaced by nothing (it represents the unmarked state), we have that um = m and mu = m. Thus m is dominant and it now makes sense that ⁊⁊ = ⁊. One can, in fact evaluate an expression by letting its deepest (empty) spaces be unmarked and then propagating state evaluations upward to the top of the expression.

For example:

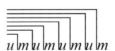

shows that

The attentive reader may remark that a nest of seven marks is marked, and indeed a nest of an odd number of marks will be marked.

We can then write algebraic expressions such as $P\overline{P}$ and ask how they will evaluate. Since P is either marked or unmarked we find here two possibilities: $u\overline{u} = \rceil$ or $\overline{m}m = \overline{\rceil}\rceil = \rceil$. Thus we can assert that $P\overline{P} = \rceil$ in the *primary algebra.*

We are now in a position to show the translation of this algebra for logic. We can write ab for "a or b" because the algebraic expression ab is marked exactly when either a is marked or b is marked or both a and b are marked. In symbolic logic one writes $a \vee b$ for "a or b." Thus we can state that $a \vee b = ab$ in the *primary* algebra. Similarly $a \wedge b$ is the standard notation for "a and b." We can write $a \wedge b = \overline{\overline{a}\overline{b}}$. Note that the only way that $\overline{\overline{a}\overline{b}}$ can be marked is if both a and b are marked! Finally it turns out that $\overline{a}b$ represents "a implies b" when we take $\rceil = True$ and $\overline{\rceil} = False$. There is more to say in this domain, but we stop here with the interpretation for logic.

On the other hand consider 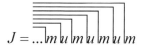.

Here we see an infinite nest of marks and we note that it appears literally underneath its own outermost mark. The infinite nest of marks re-enters its own indicational space and satisfies the equation $J = \overline{J}$. Since J has no deepest space, there is no way to evaluate it as marked or unmarked. J is neither marked nor is J unmarked. What shall we do with $J\overline{J}$? If we wish to say that $J\overline{J} = \rceil$, then we should go back to J and ask in what sense J could be marked or unmarked. Look at this state of J:

$$J = ...\overline{m}u\overline{m}u\overline{m}u\overline{m}$$

We see that it is possible to imagine J as marked, but it is also possible to imagine J as unmarked using the same procedure, labeling an infinity of spaces downward from the outside of the expression. Furthermore, if we take J as above, then

$$\overline{J} = \overline{...\overline{m}u\overline{m}u\overline{m}u\overline{m}} = ...\overline{m}u\overline{m}u\overline{m}u$$

And so we see that if J is marked then $\overline{J}$ is unmarked and indeed $J\overline{J} = \rceil$ just so long as we understand that the two J's in this equation are identical. This is the Flagg Resolution. We can incorporate the re-entering mark into the primary algebra with the understanding that it is an entity that can be either marked or unmarked and that J must be treated according to the FR.

Back to the dialogue, but now we include two new participants. Cookie and Parabel are sentient text strings who comment on all these matters from a somewhat different point of view. They are significantly closer to the void than either LK or Flagg.

Cookie: I am worrying about your infinite nest of marks. You say that if $J = \ldots\overline{\overline{\overline{\overline{}}}}$, then $J = \overline{J}$. But I have written $J\overline{J} = \ldots\overline{\overline{\overline{}}}\,\ldots\overline{\overline{\overline{\overline{}}}}$, and I see clearly that by juxtaposing J and $\overline{J}$, it is clear that $\overline{J}$ is not equal to J. It is taller by one mark!

Parabel: Do you see the three dots deep inside J and $\overline{J}$? This means that each of these expressions is infinite.

Cookie: Well, to me J is seven nested marks with three dots down in the bottom, while $\overline{J}$ is eight nested marks with three dots down in the bottom. So you see, $\overline{J}$ stands one mark taller than J.

Parabel: But adding one more mark to an infinite nest of marks does not change it.

Cookie: Why not?

Parabel. You can match them up. Its like this. Suppose I define a correspondence of the natural numbers to themselves by the function $F(n) = n + 1$.

Then $F(1) = 2$, $F(2) = 3$, … and I can make a 1 to 1 correspondence between the set $\{1,2,3,\ldots\}$ and the set $\{2,3,4,\ldots\}$.

Cookie. The way I see it is like this

$$\{1,2,3,4,5,6,7,\ldots\}$$
$$\{2,3,4,5,6,7,8,\ldots\}$$

You correspond 1 to 2, 2 to 3, 3 to 4, 4 to 5, 5 to 6, 6 to 7, 7 to 8 and the ellipsis … to itself. Easy! Each set has eight elements (letting … be one element). Nothing strange here and no infinities.

I appreciate that you can fantasize about infinities as much as a string can fantasize, but these nests of marks you have written are finite. It is just not true that you can write.

Anyone can see they are not equal. One has seven marks and three dots. The other has eight marks and three dots. Seven is not equal to eight. I may be just a string, but I know that!

Parabel: You are right. One can fantasize about a nest of marks that goes down forever, but the logical syntax of such notations is based on an equivalence relation. I can write

$$\overline{...|} = \overline{\overline{...|}}$$

and this is not a literal equality. It is the generator of an equivalence relation, a rule for substitution. And then I can write

$$\overline{...|} = \overline{\overline{...|}} = \overline{\overline{\overline{...|}}} = \overline{\overline{\overline{\overline{...|}}}} = \overline{\overline{\overline{\overline{\overline{...|}}}}}$$

by using this rule of substitution again and again.

Cookie: Oh. You are doing recursion by substitution. Every string knows about this. But wait a moment. Suppose I write the string

$$1 + x + x^2 + x^3 + \ldots$$

Then I could say that

$$1 + x + x^2 + x^3 + \ldots = 1 + x + x^2 + x^3 + x^4 + \ldots$$

In fact the recursive rule is now that

$$x^n + \ldots = x^n + x^{n+1} + \ldots.$$

It is a different rule!

You can't just substitute $\overline{...|} = \overline{\overline{...|}}$ when doing this algebra.

Parabel: You are right. We handle the ellipsis ... just as we handle a paradoxical element J in the Flagg resolution. The substitutions that are allowed for the ellipsis ... are context dependent, and we learn to behave correctly with respect to the ellipsis ... as we learn to use mathematical language.

Cookie: So the Flagg Resolution has been around all along!

Parabel: Yes it has, but James Flagg was surely the first person to become conscious of the FR. That was a giant step for mathematical consciousness.

Cookie: I have heard these peculiar words: The form re-enters its own indicational space. I think earlier you were trying to convince me that

$$\text{if } J = \overline{\overline{...|}}, \text{ then } J = \overline{J}.$$

Parabel. Yes! And now we can see this on our own grounds. You have agreed that $\overline{\underset{...}{\overline{|||}}} = \overline{\underset{...}{\overline{|||}}}$, and so, with $J = \overline{\underset{...}{\overline{|||}}}$, we have to simply substitute J into that formula and obtain $J = \overline{J|}$.

Cookie: It is not literally so unless we use the generating property of the ellipsis. I have heard that some of our readers have a fantasy that the nest of marks with an ellipsis at the bottom means an endless descent of marks. From the point of view of a string this is complete nonsense.

Nothing written goes on forever. I am glad that you explained the generative property of the ellipsis so that all this is clear to a simple string such as myself.

Parabel: I have noticed something curious. This use of the ellipsis is actually the same as the use of the reentry turn in the reentering mark.
Look at this glyph in Figure 0.

Figure 0. The Re-entering Mark.

The in-turning line indicates where the form is to be placed inside itself. To make matters clear, let me make a notation that we can use for our discussion. I will write as in Figure 0 for the Re-entering Mark using a "hat" $\wedge$ to indicate the point of re-entry.

Figure 0'. Our notation for the Re-entering Mark.

This means that we can write that if $J = \overline{\wedge|}$, then $J = \overline{J|}$, and indeed $\overline{\wedge|} = \overline{\wedge|}$. But now you see that the "hat" $\wedge$ behaves just like the ellipsis except that it is specialized for re-entry.

Cookie: That helps even more. Those readers who think that they are having infinity with the ellipsis in a really different way than they are having infinity with the re-entry symbol can now understand that there is no difference at the base of things in our string world.

Parabel. We can make other re-entry forms this way. For example, I like $F = \overline{\wedge|\wedge|}$ which implies that $F = \overline{F|F|}$. This is the *Fibonacci form* because as you keep substituting it into itself it produces an architecture with Fibonacci numbers of divisions at successive depths of the form. We can write for example:

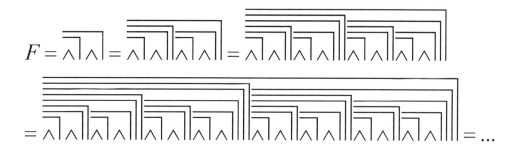

and keep going as far as we like.

Cookie: That looks like it was produced by the Mad Hatter! If you see a re-entry form, you know from its "hats" that it has to be handled with the kid gloves of the Flagg Resolution.

IV. James Flagg on the Flagg Resolution

Since the first publication of *Laws of Form*, it is well known that indication cuts deeper than truth. Truth is a token of indication. In the form of indication we can treat self-reference as a function of self-indication.

In so doing, Spencer-Brown was able to discard Russell and Whitehead's theory of types, and uncover—from within the original paradoxes themselves—expressions of higher degree, expressions concealed by the semantic, syntactical and alethic clothing of the paradoxes themselves, shorn of theory.

Varela (1975), in reviewing the application of self-indicative expressions in autonomous autopoietic domains, conceived self-indicative expressions as a "departure from the calculus of indications proper, into re-entering forms ... [as] ... not without its difficulties which render the treatment of higher degree equations, as it now stands, in need of revision" (Varela, 1975, p. 6). Spencer-Brown's claims that "it is evident that (the two algebraic initials) hold for all equations, whatever their degree" (Spencer-Brown, 1969, p. 57)

Thereafter, Varela considers the simple second-degree equation

$$F = \overline{F\vert} \qquad (1)$$

with respect to the first algebraic initial, J1

$$\overline{\overline{P}\vert P}\vert = \qquad (J1)$$

Which he rewrites as $\overline{P|P} = \overline{}|$, which is proper, since in this form "all the relevant properties of the point p in Figure 1 (Spencer-Brown, 1969, chapter 11) appear in two successive spaces of expression. In this case the superimposition of the two square waves in the outer space, one of them inverted by the cross, would add up to a continuous representation of the marked state there" as in Varela's J1.

Varela goes on to say that if Spencer-Brown is right, we have

$$\overline{}| = \overline{F}|F \quad \text{(J1)}$$

$$= FF \text{ (by 1 and substitution)}$$

$$= F \text{ (C1, C3)}.$$

But, he adds, this is clearly untrue, since replacing in (1) we obtain $\overline{\overline{}|}| = \overline{}|$ contrary to J1 itself. So, by allowing re-entry we lose connection with both the arithmetic and the algebra.

However, Varela could have remained in connection with both the arithmetic and the algebra if he had conceived $F = \overline{F}|$ (1) as a formal application of re-entry to itself. Which is to say, if an act of self-reference gets us into paradox, then a second act of self-reference reapplied to the first will get us out.

Take $F = \overline{F}|$ as a formal definition of a second act of self-reference in the form, of substitution of the (formal) definiendum for the (formal) definiens. It doesn't matter which is taken as which, so long as this substitution is performed for all appearances of J throughout a formula or not at all. This is the formal statement of the Flagg Resolution (FR).

Thus, in $F = \overline{F}|$, F has been re-defined as $\overline{F}|$, so for every occurrence of F we must substitute $\overline{F}|$ throughout, salva veritate.

Varela's J1 is $\overline{}| = \overline{F}|F$, and is transformed by FR substitution to $\overline{}| = \overline{\overline{F}|}|\overline{F}| = \overline{F}\overline{F}|$ thus preventing the collapse (by substitution) of J1 into $\overline{}| = FF$, and F.

A second pass of the resolution returns $\overline{}| = \overline{F}|F$ to $\overline{}| = \overline{F}\overline{F}| = \overline{F}|F$ which accords with Spencer-Brown's statement regarding two successive spaces of expression holding all the relevant properties of the point F.

Furthermore, from $\overline{}| = \overline{F}|F$ we may set J1 to either the marked state, as Varela did, or the unmarked state as Spencer-Brown does in as the first initial of the primary algebra. All we need do is apply the dictum of the resolution once again, what we do locally, we do globally, and from $\overline{P|P} = \overline{}|$ we obtain $\overline{P|P}| = \overline{\overline{}|}| =$. We are now back in the primary algebra, and need only recall J2 to once again fully avail ourselves of its powers.

The reader should note that the identity $F = \overline{F}|$ can be seen as asserting that any given form is identical with what it is not. This principle of identity is entirely coherent with the primary algebra via the Flagg Resolution.

The key philosophical precursor to the Flagg Resolution is Leibniz's position on intersubstitutivity: "Salva veritate (or intersubstitutivity) is the logical condition by which two expressions may be interchanged without altering the truth-value of statements in which the expressions occur" (https://en.wikipedia.org/wiki/Salva_veritate).[3]

V. Einstein/Penrose Tensor Networks and the FR.

Cookie: Are there other examples of implicit FR in ordinary mathematics?

Parabel: Here is an example. Do you remember the Einstein Summation Convention?

Cookie: That is the one thing that Einstein did that I do understand. Einstein said that we could write $G = T_{ab} S^a R^b$ and it would be understood that this quantity is obtained by summing over the instances of the repeated indices. Thus if a and b can range over the values 1 and 2, then
$$G = T_{11}S^1R^1 + = T_{12}S^1R^2 + T_{21}S^2R^1 + T_{22}S^2R^2.$$
In order for the value of G to make sense we do have some leeway in handling the indices. For example, we can write
$$T_{ab} S^a R^b = T_{ij} S^i R^j$$
but we must follow the rules:

1. Since a and b are distinct, a new variable substituted for a must be distinct from b. And a new variable substituted for b must be distinct from a.
2. If a new variable is to replace x, then it must replace all instances of x in the given formula.

It should be clear that these rules are needed in order to insure that the summation the condensed formula represents does not change when we change the indices. The rules are another version of the FR. Note particularly that the index a may appear in many places in a formula and it has the same property as our J in the first example of FR: If you change it anywhere then you must change it everywhere in the same way.

Cookie: That is a wonderful example! I have a hobby that is regarded as a bit racy by some strings. I like to rewrite Einstein expressions in diagrammatic language. For example, look at Figure 1 just below my present string. In that figure you see that T_{ab} is a circular blob with two legs labeled a and b. And S^a and S^b are each circular blobs with antennae labeled a and b. The antennae of Sa and Sb are connected directly to the legs of Tab. Thus in the diagrammatic there is one connecting line labeled a and another connecting line labeled b. It is clear that if we were to change the label a to a' it would have to be done for the whole line, and so in the algebraic expression, both a's would become a'. Here we see that the FR is working because there really is only

---

3. English translations of the definition of intersubstitutivity (salva veritate) are difficult to locate. Wikipedia translates the definition found in Leibniz's *General Science* (c. 1677; Chapter 19, Definition 1) as: "Two terms are the same (eadem) if one can be substituted for the other without altering the truth of any statement (salva veritate)" An extensive discussion of substitution can be found in Leibniz (1966).

one *a* and one *b*. In fact, in the Figure we have shown how the structure of the *T* and the two *S* blobs is actually standing independent of any particular labeling. The labeling is a way to communicate the connections in the structure of the diagram. This has many implications for numerous subjects where diagrams and interconnections are important to articulate. My favorite paper on this subject is by Roger Penrose (Penrose, 1971).

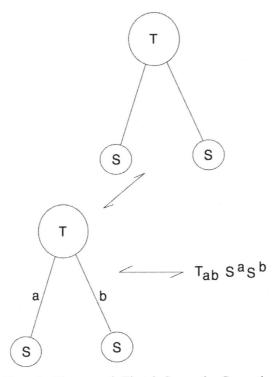

Figure 1 – Diagrammatic Einstein Summation Convention

Parabel: That is the longest speech I have ever heard you give Cookie. I can see why some strings are unhappy with your views. You have left the line for the plane. I myself am a bit leery of such excursions, but I think you are completely sound in what you say!

Cookie: I am so enthusiastic about this! Look Parabel at Figure 2.
There you see the small diagrams $Cup = M^{cd}$, $Cap = M_{ab}$ and crossings $R^{ab}{}_{cd}$, each with their own indices, and how they can be composed into the big tensor Z_K that represents a knot (Kauffman, 2012)! Just think about this. A string Z_K can precisely represent a knot in three-dimensional space. We strings are not really restricted to one dimension any more. All we need is the correct use of Flagg Resolution for the indices of our Einstein tensors that represent these higher dimensional structures. Notice that all we need, to follow the Flagg Resolution Einstein convention, is the proper use of upper and lower indices, a tradition of long standing in the string-world.

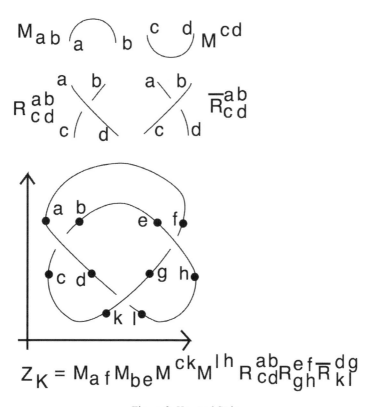

Figure 2. Knotted String

Parabel: Cookie, I believe that you have invented knotted strings!

Cookie: Not bad for a string to invent knotted strings.

VI. The Russell Paradox

Cookie: Can we look at the Russell Paradox?

Parabel: Certainly. I prefer to use string notation. Do you mind?

Cookie: Not at all. You use Sx to mean "x is a member of S." Is that what you like?

Parabel: Exactly! The Russell set is the set of all sets x that are not members of themselves. I write it as $Rx = \sim xx$, and in LOF notation I write the Russell set R as $Rx = \overline{xx}$ where it is understood that xx means "x is a member of x."

Cookie: And then we have the paradox of "Could R be a member of R?"

Parabel: In our notation we have $RR = \overline{RR}$ and so R is a member of R exactly when R is not a member of R. This is a clear case for the Flagg Resolution. We just treat RR with kid gloves and never change it except globally. With the new rule of substitution we do not have a problem here.

The Flagg resolution says: Yes we can have R a member of R or R not a member of R, but when you shift from one stance to the other it is a global shift. The two states do not occur together. There is no contradiction.

Cookie: This is very simple. How did the old time set theorists handle the problem?

Parabel: Well you know that Russell and Whitehead, in their monumental treatise *Principia Mathematica*, handled the paradox by using a theory of types. I will not go into this, but the set theorists who came after used a highly simplified type theory where there were two types of collections.

This is the Goedel, Bernays, von Neumann (GBN) *set theory* (Mendelson, 1997). In their theory a collection is either a *set* or it is a *class* (sometimes called a proper class). Sets have sets for members and every set is a member of another set. For example if S is a set then S is a member of the singleton $\{S\}$. Classes have members that are sets, but no class is a member of a set. Nor is a class a member of a class, since all members of a class are sets.

Cookie: I think I see how it works. We take the collection R of all sets that are not members of themselves. Then we ask, is R a member of R? If it were we would get the contradiction. So we conclude that R cannot be a set. Therefore R is a class. R is the class of all sets that are not members of themselves. The contradiction is gone!

Parabel: I think this GBN set theory that I have been describing to you is very like the Flagg Resolution and yet very different! One determines that R is not a set and hence a class. Once R is a class, we cannot form $\{R\}$ or other ways to treat R as anything but singular. It is not the same as changing all instances of R at once. We could try an FR where we can have RR or $\sim RR$ but if we write $T = RR \vee \sim RR$ we cannot substitute RR for $\sim RR$ and conclude that $T = RR \vee RR = RR$ and then conclude that $T = \sim T$. This will be forbidden directly by the FR. The GBN resolution by sets and classes is a separate move that removes the paradoxical element completely. It is like putting an imaginary value in a circuit transition to keep it from being oscillatory.

Cookie: Can you explain that?
Parabel: Why yes I can. Consider the circuit in Figure 3

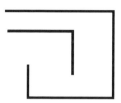

Figure 3. A Two-State Circuit

If the values in the two inner spaces of this form are both unmarked and we let the circuit perform, then if the inner mark fires first we get an unmarked state on the outside and a stable form, while if the outer mark fires first we get a marked state on the outside and a stable form. This re-entrant form has an ambiguous transition to its next stable state. In the case of the circuit it is possible to add an extra feedback that influences this transition and makes it determinate. This idea is analogous to replacing $Rx = \overline{xx}$ with $Rx = \overline{xx} \wedge Sx$ where $\wedge$ denotes "and." R now states that its members are not members of themselves and they are sets.

Cookie: I could rewrite it as

$$Rx = \overline{xx} \wedge Sx = \overline{\overline{xx}\|\overline{Sx}\|} = \overline{xx\,Sx\|}$$

Thus $Rx = \overline{xx\,Sx\|}$.

Then

$$RR = \overline{RR\,SR\|}$$

and if SR is unmarked (R is not a set, hence a class) then $RR = \overline{RR\,\|} = \overline{\|}$ and so it is simply the case that R is not a member of itself when R is not a set. If R is a set we obtain a classical contradiction. Since we assume in this classical set theory that it is free of contradiction, it follows that R is a class and not a set. R is a class and cannot be a member of any set or class. The paradox is resolved by the extra observer in the circuit. We simply do not get to consider the paradoxical expression $RR = \overline{RR}$. This is how the Goedel, Bernays, von Neumann set theory resolves the Russell paradox. It uses the types of set and class to banish the paradoxical statement from every appearing in the language of the theory.

Parabel: We could have done this to resolve any paradoxes. If we have $P = \overline{P}$ we replace it by $P = \overline{P\,SP\|}$ and then SP =True gives a contradiction, so SP must be false. That throws P in the $\overline{S\|}$ category and essentially into jail! Do not pass Go, go directly to Jail. This is the GBN solution placed in the simple context that a string can understand.

Cookie: The GBN seems very artificial from the point of view of a simple string. The notion of simultaneous replacement in FR is an exciting new dimension for a string.

Parabel: One small step for a string. But a giant step for stringkind. I hope you are not stringing me along with this flattery.

Cookie: On that note we have to vanish into the void. But maybe we can reappear in another publication.

Parabel: Where else do you find strings of symbols? It is publish or perish for us. Either we are in print or we are not. Cookie?

Cookie:

Parabel: Gone again into the void.

VII. Epilogue

This essay on the Flagg Resolution has been an essay on the role of substitution in mathematics. When we write A=B where A and B are in fact different in some way, then one must take care to see just what it is that makes them equal and what it is that makes them different. When we have an apparent logical paradox in the form $J = \overline{J}$, the most careful way to handle the substitution is to regard all appearances of J in a given expression as identical and let them change together or not at all. When we break a whole into parts and give remote labels on the parts to enable the reconstruction of the whole, these multiple labels for the same place must be regarded as identical. They can be changed together or not at all.

In this way the whole and its parts can be identical in the form. Unity and diversity can coexist, and circularity can take its proper place in logic, reason and structural understanding.

References

De Nicolas, A. T. (1978). *Meditations through the Rg Veda—Four-Dimensional Man.* York Beach, ME: Samuel Weiser, Inc.
Kauffman, L. H. (1999). Virtual logic – The Flagg resolution. *Cybernetics and Human Knowing, 6* (1), 87–96.
Kauffman, L. H. (2012). *Knots and physics* (4th ed.) Singapore: World Scientific.
Kauffman, L. H. (2017). Virtual logic – Cookie and Parabel discuss laws of form. *Cybernetics and Human Knowing, 24* (3-4), 261–268.
Leibniz, G. (1966). *Logical papers* (G. H. R. Parkinson, Ed. & Trans.). London: Oxford University Press.
Mendelson, E. (1997). *An introduction to mathematical logic* (4th ed.). London: Chapman and Hall/CRC.
Penrose, R. (1971). On applications of negative dimensional tensors. In *Combinatorial mathematics and its applications* (221–244). London: Academic Press.
Spencer-Brown, G. (1969). *Laws of form.* London: George Allen and Unwin Ltd.
Varela, F. J. (1975). A calculus for self-reference. *International Journal of General Systems, 2*(1), 5–24.
Waismann, F., & McGuinness, B. F. (1967). *Wittgenstein und der Wiener Kreis.* Oxford, UK: Basil Blackwell.

Paper Computers, Imaginary Values and the Emergence of Fermions

Louis H. Kauffman[1]

This paper shows how, when Boolean algebra turns upon itself and becomes self-referential, the elements of oscillation, memory, counting, imaginary values and the beginnings of quantum physics in the form of Fermions and the Dirac equation are born.
Key Words: laws of form, marked state, unmarked state, self reference, Boolean algebra, Boolean arithmetic, circuit, memory, oscillator, modulator, time, synchronous, asynchronous, imaginary value, iterant, iterant algebra, Clifford algebra, Fermion algebra, relativity, Dirac equation.

I. Introduction

There is a well-known and fundamental connection between Boolean algebra and switching circuits. First formalized by Claude Shannon (Shannon, 1938), this connection has formed the basis for the development of computer circuitry, and it underlies many subsequent developments in programming, machine design and mathematical logic. The purpose of this paper is to put forth an elementary vision of this connection that goes further than the usual approach. I wish to show, using very simple mathematical models, how, when Boolean algebra turns upon itself, when it becomes self referential, then memory, counting and indeed the world of Fermions and their algebra are born! This insight is implicit in the work of G. Spencer-Brown in Chapter 11 of his work *Laws of Form* (Spencer-Brown, 1969), but we develop the idea ab initio here. We will do more and show how certain basic aspects of quantum physics related to Fermions and the Dirac equation emerge from these structures.

Some words about notation are appropriate. In *Laws of Form*, Spencer-Brown uses the mark ⏐‾ and builds his notation on that basis. The mark is seen to make a distinction, as though it were a box ☐, between its inside and its outside. In this sense, the mark is self referential, and can itself be regarded as an imaginary value. The calculus for the mark is based on two equations, the law of calling and the law of crossing:

$$\overline{\overline{}}\,\, = \overline{}$$

$$\overline{\overline{}} =$$

---

1. Mathematics Department University of Illinois at Chicago; Department of Mechanics and Mathematics Novosibirsk State University, Russia. Email: kauffman@uic.edu

In the first equation, two marks each call the name (marked) of the other mark. The call is redundant and can be replaced by one of them.

In the second equation, the mark acts as an operator to cross from the marked state indicated on its inside to the unmarked state. The two nested marks are replaced by an unmarked state (the blank notational plane). The system of expressions involving the mark is two valued. Any expression can be uniquely reduced to either the marked or the unmarked state. Algebra arises from this notation as in the equation $\overline{\overline{a}|}\,| = a$. This equation is seen to be true, since if a is unmarked it is identical with the law of crossing above, and if a is marked then we have $\overline{\overline{\neg}|}\,| = \neg|$, again by the law of crossing. The mark is seen to be both a value and an operator in this Calculus of Indications. In the algebra, the mark as operator is the analogue of negation in logic or Boolean algebra. It is often useful to use other typographical notations for the mark. Thus we shall often write $<> = \neg|$ and $<a> = \overline{a}|$.

And so we have the Laws of calling and crossing in the forms $<><> = <>$ and $<<>> = $. Calculi written in forms analogous to the Spencer-Brown calculus will be called containment calculi, since they utilize the distinction making properties of the notations in which they are composed. For those who enjoy this work, there is something charming about taking the advice of ones own notation in the composition of mathematical ideas.

The reader should consult the section headings of the paper for a guide to its contents. We begin with a fable about the interactions of self-referential distinguishers that often disappear due to the enjoinders of the Pauli exclusion principle in a universe so early that it scarcely has but one distinction. We then get down to business and show how the distinguishers pair up and form memories, and then they form modulators that can divide the frequency of an oscillation by two. The modulators are asynchronous, working independently of any particular time delays that might happen in later universes where time had appeared. The discipline of designing correct asynchronous modulators occupies us, and we develop various approaches to it. In the course of this modulator discipline we discover how imaginary values make the smallest modulators operate correctly. The imaginary values are states of the circuit that last just long enough to influence a transition, and then disappear to a stable real value. We then explore imaginary values and the transitions of the reentering mark, a simplest circuit J that oscillates always because it feeds back to itself the opposite of its present value.

$$J = \overline{J}|$$

Such a circuit is a pure imaginary. We then find that there are arithmetics and algebras associated with the oscillation of the reentering mark (and its numerical analog—the oscillations for the square root of negative unity). We discover that an oscillation between plus one and minus one leads to the square root of negative unity and to Clifford algebra, while the oscillation of 0 and 1 (Boolean or numerical) leads to the creation and annihilation algebra for a Fermi particle.

It is worth pausing in this summary to give a quick picture of the relationship of a zero-one oscillation and the Fermion algebra. See sections VII to XII for the details.

Start with the basic oscillation

$$...010101010101....$$

The 0 and the 1 can be Boolean or numerical.
We represent a waveform

$$...ababababab...$$

by two ordered pairs (a,b) and (b,a). They indicate two ways to view the oscillation as starting with a or starting with b. To combine waveforms we use the rules

$$(a,b)(c,d) = (ac,bd) \text{ and } (a,b)+(c,d) = (a+c, b+d).$$

We call the ordered pairs *iterants*.
The two iterant views of the 0-1 oscillation are:

$$p = (0,1) \text{ and } q = (1,0).$$

We see that

$$pq = 0$$
$$p+q = 1$$
$$pp = p$$
$$qq = q.$$

We use a *time shifter* η with

$$(a,b)\,\eta = \eta\,(b,a).$$

In Section X we show that this property of the time shifter η follows from the existence of *time sensitive entities* with natural properties. Note that

$$\eta(a,b)\,\eta = (b,a) \text{ and } \eta\,\eta = 1.$$

Let $U = p\eta$ and $U^{\dagger} = q\eta$. These are the time sensitive iterants. Time sensitive iterants interact in time and in space. For example $UU^{\dagger} = p\eta q\eta = pp = p$, In this interaction q is shifted by one time step, becomes p and interacts with p to produce p.
It is easy to check that

$$U^2 = (U^{\dagger})^2 = 0,$$

$$UU^{\dagger} + U^{\dagger}U = 1.$$

This is the creation/annihilation algebra for a Fermionic particle such as an electron.

The two sequences ...0101010101... and ...+1,-1,+1,-1,+1,-1,... have very different meanings. The zero-one sequence is primordial. It goes back to the presence and absence of a simple distinction. The plus-minus sequence uses numerical structures that come later in the development of structures and ideas. In standard arithmetic, we take minus as analogous to negation. But the system with +1 and -1 is quite different from the zero-one system.

0 is self referential in the + - system! After all, $-(+1) = -1$ and $-(-1) = +1$ but

$$-0 = 0.$$

Zero is invariant under numerical negation.
Zero is that place where the positive and the negative come together.
Zero is self referential.

The plus-minus oscillation ...+1,-1,+1,-1,+1,-1,... leads to the square root of negation and Clifford algebra and Fermions as well, but the zero-one oscillation ...0101010... is fundamentally Fermionic. Boolean algebra turns upon itself and produces the beginnings of quantum physics.

In Section XI we find a non-commutative containment calculus for the Fermion, and in Section XII we show a Pythagorean relationship that binds a nilpotent Clifford algebra element to the Dirac equation. A nilpotent algebra element U has the property that its square is zero: $U^2 = 0$. This Fermionic rule is the algebraic counterpart to the law of crossing

$$\overline{\overline{}} = .$$

Crossing states a fundamental law of form –

> A distinction, fitting into itself,
> Recognizes that what it is,
> Is identical with
> What is not,
> And the distinction
> Disappears.

The laws of appearance and disappearance in the Universe are the precursors to the laws of physics that we find for the apparent appearances and disappearances and patterns of existence in the world of measurement and experiment. Thus it is natural that physical law should be seen as a generalization of the laws of calling and crossing in a calculus of indications. At the end of the essay we show how the nilpotent algebraic elements that arise from a structure of oscillation and imaginary value give rise to the nilpotent solutions to the Dirac equation and indeed to the Dirac equation itself. This aligns our story with the work of Peter Rowlands (2007). The Dirac equation is the fundamental relativistically invariant equation incorporating the spin of the electron. This journey from the imagination of a distinction to the structure of mathematical physics is only a beginning. There is more to say and there will always be more to say for those who have some luck in the articulation of what a distinction would be if there could be a distinction.

II. A Beginning

Lest these statements seem excessively esoteric, lets look at once at the case of memory: Imagine building a machine from basic elements that invert a signal. There are two signal types labeled 0 and 1 and satisfying

$$00 = 0$$
$$01 = 10 = 0$$
$$11 = 1$$

Thus the 0 signal is dominant. The equation $01 = 0$ denotes the fact that the one (1) is dominated by the presence of the zero (0). We denote an inverting element by a diagram in the form of a line with a bold arrow in the middle. Signals that pass the arrow are inverted.

The signals travel along the line in the direction of the arrowhead. Thus we have

as indicators of the process of inversion.

Let $<a> = \overline{a}$ denote the result of inverting a. Thus $<0> = 1$ and $<1> = 0$. Then we can write

to indicate the process of inverting a.

Since we also have a notion of signal combination ab, this also deserves a diagram

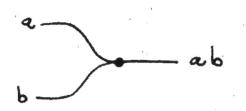

We will use the typographical notation of Spencer-Brown's *Laws of Form* (Spencer-Brown, 1969) for the inversion so that <0> = 1 and <1> = 0. Note that we then have a number of algebraic rules such as <<a>> = a for any a, and <<a>a> = 1 for any a. Each such identity can be verified by checking it at 0 and checking it at 1. Thus <<0>> = <1> = 0, <<1>> = <0> = 1 so in all cases <<a>> = a.

Spencer-Brown noted that one can take <> = ⌐ to stand for 0, and a blank space (the unmarked state) to stand for 1. This use of the unmarked state is conceptually of great value, and it means that the calculus in laws of form is generated by the one symbol ⌐. Thus in laws of form one writes

$$<0> = 1 \text{ as } <\diamond> = \quad ,$$
(since 1 is an empty space)
and
$$00 = 0 \text{ as } \diamond\diamond = \diamond.$$

These two equations
$$<\diamond> =$$
$$\diamond\diamond = \diamond$$

become the basis of the calculus of indications of laws of form. Incidentally, the other Boolean operation, a + b, can be defined by a+b = <<a><b>> and so is also represented by a diagram.

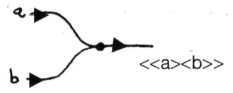

Note that <<a><b>> = 0 if and only if a = 0 and b = 0. This diagram represents the logical operation of "and".

Now comes self-reference! Consider the equation

$$(*) \quad x = \overline{x}\rceil.$$

This is the Boolean analog of the liar paradox, "This statement is false." For if we assume that x=0, then (*) implies that x = 1 and if we assume that x = 1, then (*) implies that x = 0. So x must be neither 0 nor 1! Can it be both? Well maybe, but let's diagram the equation (*):

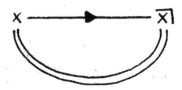

You see that $x = \overline{x}$ says that the output of the inverter has been turned back and plugged into the input:

Here the self-reference has become a circularity and we suddenly get a new view of the paradox. Such a circuit will simply oscillate!

$$...0101010101010101...$$

That is, if you actually make such a device, there will be a little time delay as the signal passes through the inverter. During this time interval the output remains 0, then flips to 1, then flips to 0, and so on ad infinitum.

There has been a bit of mathematico-linguistic sleight-of-hand here. Consider the concept of time-delay as associated with physical inverting elements. By interpreting this framework, the sense of paradox has shifted into the understanding of the possibility of temporal oscillation. This is quite legitimate.

On to memory: Look at this circuit.

Here y is the output of the left inverter, and x is the output of the right inverter. Thus $x = \overline{y}$ and $y = \overline{x}$. Circularity again. But now there is no oscillation. If y = 0, then x = 1 (and if x = 1 then y = 0). On the other hand, if y = 1 then x = 0. The circuit has two stable states:

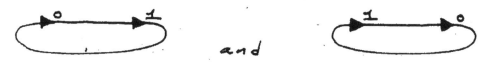

This is an entirely new and unprecedented phenomenon. By allowing apparently paradoxical circularity in our descriptions, we have produced the design of simple machines that can maintain themselves in more than one state of existence. These stable states have nothing to do with time delays. The left inverter steadily gives its 0 signal to the right inverter who, in turn, steadily provides a 1 to the left. They hold each other up in the mutual embrace that is the stable state. Engineers call this the steady state behavior. It is adequately modeled, as we have done, by asking that the graph of inverters and lines be labeled so that each inverter is balanced with the value on its right the opposite of the value on its left.

But how did this stable state arise? Well first of all the two inverters had to get together. This is a matter of conjecture and mythology, but I sometimes imagine the following scenario that begins the next section.

III. Long Ago and Far Away

Long ago and far away, there was a primeval soup of curled up single self-referential inverters, each oscillating away, ignoring all the others. (Perhaps they were a product of the Big Bang.)

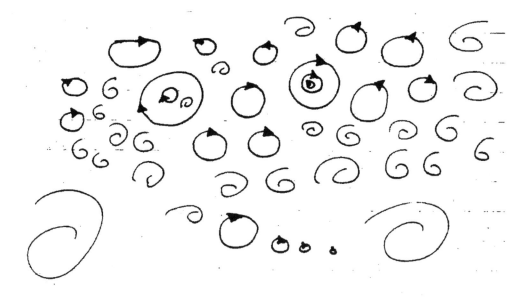

These were very primitive times indeed and in fact the only "things" around that could be used as signals were the little curls themselves! And there were not any 1's or 0's. Such names came much later. But that was alright, since an inverter would just merrily produce a primordial zero

$$(\; o \; (\; \eth \;) \;)$$

in the form of a self-returning curl with no inverter (a self-reference of the void) if nothing came to its input, and it would stop if something came along:

If it were not for that sense of direction given to each inverter, this might seem a bit confusing. But that directionality was the legacy of the Big Bang.

Oh! I did not tell you how this works! It went this way. In order for an inverter

to act on the curl

and cancel it, the inverter would curl around its victim and engulf it right properly:

Then it would move right down and superimpose itself on the unsuspecting curl.

And due to a very ancient universal law (to the effect that you cannot have two different things in the same place (Wolfgang Pauli's exclusion principle) the two curls would just vanish!

Getting something from nothing is a bit harder, but I am told that it was entirely due to the inverters tendency to curl on themselves. Of course nothing can do only nothing, and so nothing curls on itself and so you have something.

Ah, but I digress. Back to the memory. Consider the following scene. Two curls find themselves in close proximity (this initial high temperature of the big bang has cooled considerably). They meet and touch along part of their self-referential boundaries. These cancel Pauli-wise and lo! An interlocked pair is born.

This is only the beginning of a long and complex history. Unfortunately, due to enormous cancellations, there are very few further references or records of this evolution. I have it on good authority however that the Big Bang itself was the result of a single self-reference. In those ancient epicyclic days of wheels on wheels on wheels, there was much to be grateful for in the idemposition law (as the Pauli principle was later called by George Spencer-Brown [Kauffman, 1985]) that allowed even a little cancellation.

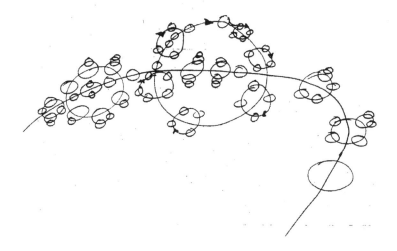

Returning to the mathematics of memory, we can imagine the stable state being induced by holding one input at 0 until the memory flips to the appropriate state.

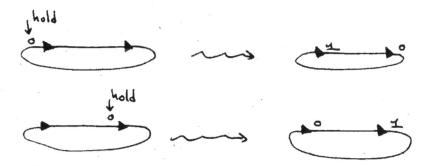

If the zero is held long enough then the signal gets through both inverters, completing the cycle by itself and the need for outside assistance is gone. The state is remembered. Since the circuit can be influenced into either of its two states, it becomes a repository for a single bit of information.

In the next section we shall carry out this story further and examine the design of circuits that can count and do arithmetic operations.

IV. Memory and Flip-Flops.

Lets examine the memory circuit M further. We allow inputs a and b so that

$M:$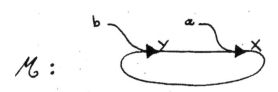

$x = \overline{ay|}$ and $y = \overline{bx|}$.

Suppose that a =1, b=1, y =0, x = 1.

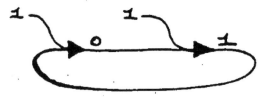

This is a stable state for M. Now let b change to 0 (we write b => 0).

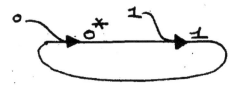

The * indicates that the circuit is unbalanced at y. Since b is being held at 0, y => 1.

Now x => 0:

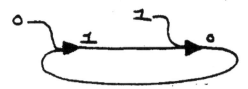

This is the new stable state.

Now note: In this state the circuit is opaque to changes in the input b. We must change a to cause another transition. This is an important characteristic of this memory circuit. It has two inputs and two states. For each state only one input is sensitized. Flipping states also flips input sensitivity.

Lets put the situation in another frame. Our memory circuit is analogous to a box with two buttons (A and B) and two lights (Red and Green). At any given time

either Red or Green is on. If Red is on, then a push on A will switch the box to Green. Further pushing on A will not have any effect. To change back to Red, you must push the B button. A second push on B has no effect.

The remarkable fact is that we can use this very insensitivity of the memory to counting beyond two in the design of circuits that can actually count! The basic idea is a device called the flip-flop for which the best description I know involves people. Here are the rules:

1. Your right arm is either Up or Down.
2. When your left palm is touched, change the position of your right arm.
3. When your left palm is released, do not change the position of your right arm.

Note that under repeated release and touch (R and T) the up-down (U and D) behavior occurs with half the frequency of the initial release and touch. Thus the flip-flop divides by two.

Now it is possible to use our memory box to produce this behavior. We just need to alternate the operation of the buttons!

```
Touch A      Green
Release A    Green
Touch B      Red
Release B    Red
Touch A      Green
Release A    Green
Touch B      Red
Release B    Red
```

Now this suggests that if we could somehow surround the memory with a decision apparatus that would alternately direct a signal to one side and then to the other, we would be able to create a circuit with flip-flop behavior.

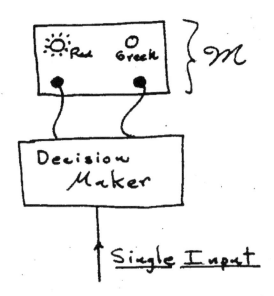

However, if this decision-maker does not make use of the information in the memory M itself, then it is as good as a flip-flop by itself. Thus we should expect a more intimate connection between M and the signal switching apparatus. The picture would be more like the diagram below.

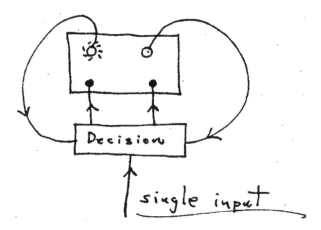

Now however we do design the decision maker it will be symmetrical with respect to each side of the memory.

For a given side there must be a part of the decision-maker that has two states: one that lets the signal pass, one that inhibits it. This suggests using another memory! And if we need one extra memory, then by the above reasoning we shall need two of them!

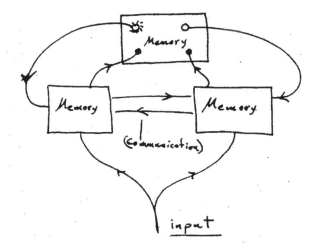

Thus now our projected design must appear as above. Presumably there may be need of communication lines between the two memories.

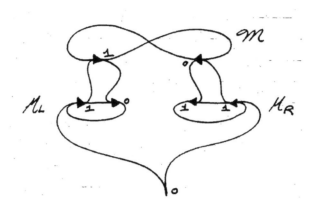

Here is a first stage attempt to implement our ideas. The main memory is represented by the large infinity form. The two auxiliary memories are represented by the two elliptical forms. In the state illustrated, the left memory is opaque to a change of $0 \Rightarrow 1$, but the right memory will change (and note that this is due to the fact that right now both its sides are held at 1: one side by the input and one side by the state of M (the big infinity sign memory).

Lets change the input $0 \Rightarrow 1$ and watch what happens.

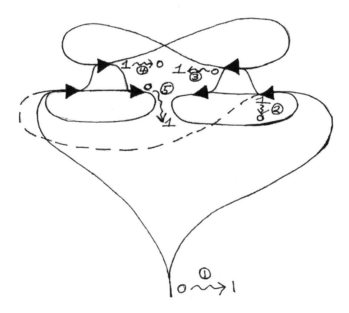

After these five changes the main memory has flipped over as desired, but the situation is still unstable. We would like a nice 0 to hold M in place! Well, a zero is available from one side of the right-hand memory M_R! See the dotted line in the diagram above.

Thus the situation suggests that we should add this communication line and (of course) its mirror image, to find a next guess for the flip-flop.

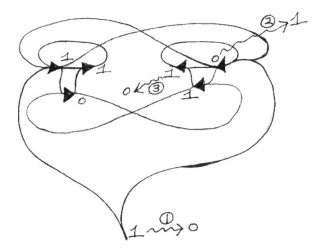

Here we have redone the drawing and labeled it with our last stable state. If the input 1 now changes 1=>0, then only M_R changes as the transitions on the drawing above indicate. Hence by symmetry (this last state is the mirror image of the state in which we began) this circuit is a flip-flop.

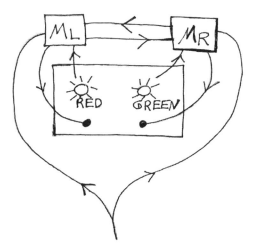

The final circuit conforms precisely to our general design notion, and it is remarkably simple in its operation. The reader should check that the circuit does indeed behave as advertised!

Remark: We should remark that the circuit that we have constructed can be described by a set of (self-referential) equations. In this case the equations are

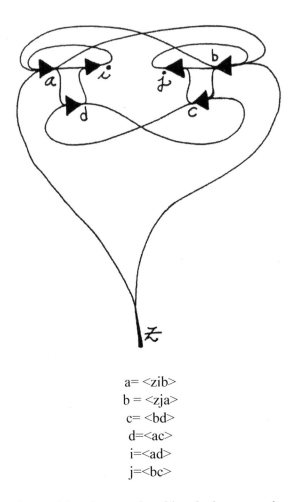

a= <zib>
b = <zja>
c= <bd>
d=<ac>
i=<ad>
j=<bc>

In order to run the model at the equational level, choose a value for the input z and then choose values for {a,b,c,d,i,j} so that all the equations are satisfied. That is an initial stable state for the network. Choose an ordering of the variables {a,b,c,d,i,j}.

Change the value of z and keep z at the new value. Go through the equations in the order chosen and when you meet the first value for which the equation is no longer satisfied (the first unbalance), change the value of that variable. Then rewrite all the other equations using that new value for (for example a). Then again look for the first equation that is unbalanced and repeat this process. Keep going in this fashion until all the equations are balanced. This is the new stable state. Now change the value of z again and go through the stabilization process again. Repeating that process with this circuit will result in an order four pattern as you change z four times. At the fourth change you will be back to the beginning. Circuits that do not depend upon the choice of ordering of these reactions are special and form the foundation of the structure of digital computers. This is a entry into asynchronous circuit design.

In the next section we will go through some of this material again with a method of working with it that you can do on paper, with markers such as Othello chips or Go stones.

Remark on counting: We end this section with an indication of how binary counting can be obtained by connecting flip-flops in series with one another. We have a cartoon human model for a flip-flop. Let us make a human counting circuit.

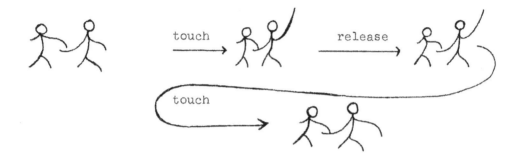

Each person becomes a binary frequency divider. If we take the *output* behavior of one person as the *input* to another person, then each successive individual will wave his right (output) arm half as fast as the mover on his left. Here is what might happen for four protagonists. (The left-most person drives the assembly.)

In this transition, A touches B causing B to touch C, and C raises his/her arm.

Here A touches B, making B touch C, making C touch D and D raises arm.

<u>Finis</u>

. . .

Notice how this chorus line of human flip-flops is indeed performing a binary count. Let 0 denote UP and 1 denote DOWN. Then the pattern of digits in the cartoons above is (for the right hand three people from left to right).

$$000$$
$$100$$
$$010$$
$$110$$
$$001$$

Rewrite this from right to left and you get:

$$
\begin{array}{c}
000 \\
001 \\
010 \\
011 \\
100
\end{array}
$$

This is the familiar binary counting sequence. It was inverted in our cartoon!

In a longer binary addition sequence the pattern becomes clear. The rightmost column oscillates with period 2, the next column has period 4, the next period 8, and so on. By connecting flip-flops together in a linear order we use their capacity to divide frequency by two, to multiply period by two and we reproduce the pattern, in detail, of binary addition. Binary addition is a property of nature—where nature is the world of self-referential entities evolving to memories and by their very nature forming modulators and combinations of modulators.

$$
\begin{array}{c}
0000 \\
0001 \\
0010 \\
0011 \\
0100 \\
0101 \\
0110 \\
0111 \\
1000
\end{array}
$$

The upshot is that once you have modulators—human or via digital circuits or other devices that can make distinctions—counting and all the intricacies of arithmetic can be performed. This is the basis for the design of digital computers, but it is deeper than that. It is the basis of the generation of arithmetic and all the distinctions and structures that are available to us when arithmetic has appeared.

V. Black and White Markers

In this section we will use the circuit notation of G. Spencer-Brown as in chapter 11 of his book *Laws of Form*. Arrows that we have formerly used as inverters are now allowed to be just arrows, pointers. And we use a vertical line segment crossing another line as the inverter. See below for many examples.

In this new notation we will use black dots for the marked state (same as 0) and white dots for the unmarked state (same as 1). The inverter is a vertical slash and signals proceed from left to right only.

Thus

$$\overline{Marked} = Unmarked$$

is depicted by

and

$$\overline{Unmarked} = Marked$$

is depicted by

The mark can be construed as an operator that inverts a signal.

$$a \longrightarrow \!\!\mid\!\! \longrightarrow \overline{a}$$

More generally, we can consider a collection of inputs to the inverter, and a collection of outputs. *Balance* at the given inverter will mean that each output is the cross of the juxtaposition of the inputs. This is illustrated below for three inputs a,b,c and one output.

$$a, b, c \longrightarrow \!\!\mid\!\! \longrightarrow O = \overline{abc}$$

The reentering mark can be construed as a circuit in which an inverter feeds back its output directly to the input. The result of such an interconnection is an oscillation between the initial state and its inversion. Such an interpretation assumes that there is a time delay between the production of the output and the processing of the input. If there is no time delay, then we are in a state of eternal contradiction.

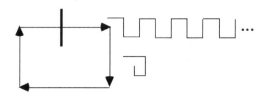

From the mathematical point of view, time is just another structure. Thus we can say

$$J_{t+dt} = \overline{J_t}$$

and as long as dt is non-zero, then there is no contradiction.

If dt = 0, then we arrive at the abstract structure of the re-entering mark that is neither marked not unmarked, the imaginary value.

There is a way to understand such circuits that goes beyond simple temporal recursion. A given circuit may have stable states, where the equations at each inverter are balanced. Then one can consider the process corresponding to the circuit to be a pattern of transitions from one stable state to another, instigated by an imbalance at some places in the circuit. If the circuit has no stable states (as with the reentering mark) then the process of transition continues without end. A transition process happens as follows:

Transition Model for Circuits.
Assign time delays to each inverter in the circuit. For the purpose of this model, it is sufficient to just order the inverters so that one can answer the question whether any one inverter is slower than another.

1. Find an inverter in the circuit whose equation is not balanced. Readjust the outputs of this inverter so that it is balanced. If there is more than one unbalanced inverter, choose the one with the smallest time delay.
2. Examine the circuit once more. If it is balanced, stop. If there is an unbalanced inverter, perform step 1 again.

Another way to formulate this model is to replace ordering of the marks by a probabilistic choice. The rules would then read:

Probabilistic Transition Model for Inverter Circuits
1. If there is more than one unbalanced inverter, choose one of them at random. Readjust the outputs of this inverter so that it is balanced.
2. Examine the circuit once more. If it is balanced, stop. If there is an unbalanced inverter, perform step 1 again.

A circuit is said to be determined if the process described above does not depend upon the (time delay) ordering of the marks in the circuit (or upon the probabilities in the second model). This transition model for circuit behavior is asynchronous in the sense that we do not assume that there is an external clock that causes all rebalancings to happen at once. Clocked behavior can be quite different from unclocked behavior.

The next example is the simplest circuit with a stable state.

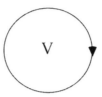

You might call this circuit V, the reentering void. It has two stable states, marked or unmarked, and no inverters. In illustrating the states, marks in the state are indicated by black dots and white dots. A black dot is a mark, and a white dot denotes the absence of the mark.

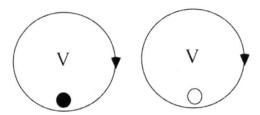

You can think of V as a form of memory, where a given state labeling persists for as long as necessary. We will, however, not use this memory, but rather the next one (see below) in making circuit designs. For mathematical purposes one could use V in circuit design, but the memory we are about to construct, by taking two inverters back to back, actually corresponds to what is done in engineering practice. At the least, if we used V, we would have to assign a time delay to it and then it would have a similar mathematical effect as the back-to-back inverters that we are about to discuss, the only problem being how to kick it out of the marked state once a mark had begun to circulate round its basic turn. The difference between V and the circuit we are about to discuss is the difference (operational at best) between

and the void.
The next example corresponds to the equation

$$M = \overline{\overline{M}}$$

and its corresponding circuit.

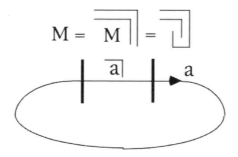

Here we have a benign reentry that does not create oscillation. The circuit has two stable states, and it is described by two equations with the extra variable N corresponding to the internal line in the circuit.

$$M = \overline{N}$$
$$N = \overline{M}$$

One need not think of any recursion going on in the stable state. In that condition, one just has a solution to the above equations. Each part of the circuit balances the other part. The circuit itself can be interpreted as a memory element, in that it can store the information time-independently.

(M,N) = (marked, unmarked)
or
(M, N) = (unmarked, marked).

These two stable states of the memory are depicted in the figure below.

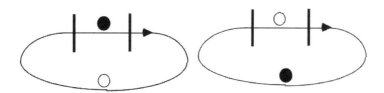

A little modification of this memory circuit, and we can interrogate it and change it from one state to another.

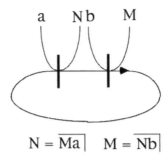

$$N = \overline{Ma} \quad M = \overline{Nb}$$

The new circuit has inputs a and b, and outputs that can measure the values of M and N without affecting the balance of the circuit itself. By choosing a marked and b unmarked, the memory is forced into the state

(M,N) = (marked, unmarked).

By choosing a unmarked and b marked the memory is forced into the state

(M,N) = (unmarked, marked).

In each case, since these states are stable, the marked input can be removed without affecting the state of the memory.

A more diabolical setting would be to have both a and b marked and then to remove them simultaneously. The resulting state of the memory is then the unstable configuration shown below.

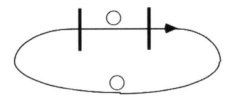

Each mark in the memory is unbalanced. The first mark to transmit a marked state will win the race and propel the memory into one of its two stable states. If it is possible for both marks to fire at once, we would arrive at the other unstable state where both sides are marked. In physical practice this will never happen, and the above unstable state will fall to one or the other of the two stable states, just as it does in our transition model (where one mark reacts faster than the other).

In practice, memory conditions such as the above can occur, and it is interesting to see how to design a circuit that will determinably transit to only one of the two possibilities. Consider the circuit below.

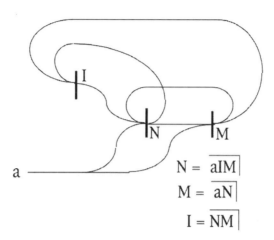

$$N = \overline{aIM}$$
$$M = \overline{aN}$$
$$I = \overline{NM}$$

In this circuit we have eliminated the arrows that indicate direction of signals through the inverters and have used the convention that signals travel through each inverter from left to right. This suffices to fix all other directed lines. The memory consisting of M and N has an input a, and there is one more mark in the circuit labeled I. This new mark observes the values M and N. If a is marked, then N and M are unmarked, forcing I to be marked. This is a stable condition of the circuit so long as a is held in the marked state.

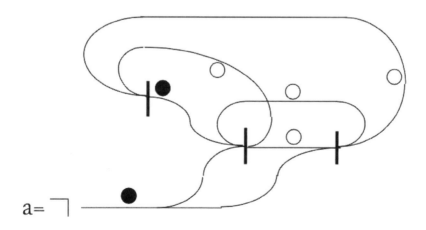

If now, we let a change to the unmarked state, then the circuit becomes unbalanced at M only, since I continues to put out a marked state.

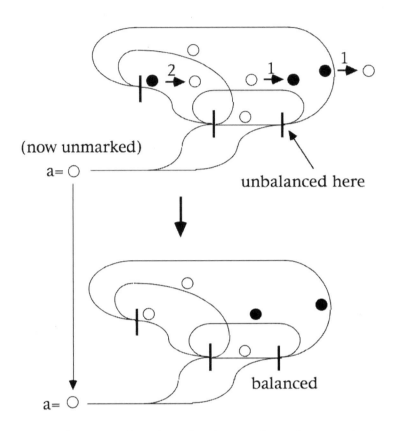

In the diagram above, we show how the change of a to the unmarked state gives rise to a transition of M to the marked state (1) and that this forces a transition of I to the unmarked state (2). The resulting circuit is balanced and the transition to this state of the memory is determinate.

The addition of the mark I to the circuit enabled the determinate transition. In fact, I acts as an observer of M and N who feeds back I = <MN> to the input to N. The result is that I is marked at the point of transition, holding the state at N in balance. For any choice of time delays, this condition of I can happen across an arbitrarily small time. But that time is significant in forcing the transition.

We see that a circuit can be construed as a miniature self-observing system, and that this condition of self-observation can radically influence the behavior of the circuit. In a certain sense, the value of the circuit at I is imaginary in at least the metaphorical sense of the term. It is "eyemaginary." In terms of circuit design, we can use such imaginary values to influence the structure of the design and make otherwise indeterminate circuits determinate. We say that a marker in a circuit has an imaginary value if there are transient states at that marker that influence the transition behavior of the circuit.

The circuit in the previous section is an example. It is a modulator in the sense of Spencer-Brown. A given frequency of waveform input at a results in an output waveform of one-half the frequency at the input. I discovered this circuit in 1978 when studying laws of form.

It is similar to circuits in chapter 11 of *Laws of Form*, and it accomplishes the modulation without using any imaginary values using only six markers. Spencer-Brown gives an example in chapter 11 of a modulator with six markers that uses imaginary values. The reader will enjoy making the comparison.

Modulators are the building blocks for circuits that count and are often called *flip-flops* in the engineering literature. There is much more to say about this circuit structure and its relationships with computer design, information and cybernetics, but we shall stop here, only to note that this is an aspect of laws of form that goes far beyond traditional Boolean algebra, and is well-worth studying and working with as a research subject.

Below we illustrate a circuit that is topologically equivalent to the one we have discussed earlier in this paper.

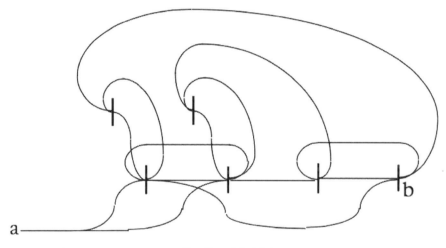

Kauffman Modulator

Remark: The reader will enjoy analyzing the next circuit. It is also a determined modulator, and has the same number of markers as the diagram above, but fewer connecting lines than the one we have constructed in the body of this paper. This design is due to G. Spencer-Brown (personal communication in 1992) and it is conjectured to be a minimal modulator, where one wishes to minimize the number of markers and the number of lines. I discovered that this design is actually used in certain Motorola flip-flops (undoubtedly independently found by chip designers using somewhat different language). The question of classification of minimal modulators is a good example of the open nature of the mathematics of circuit design. This abstract version and mathematical model of circuit design makes it possible to formulate such questions with precision.

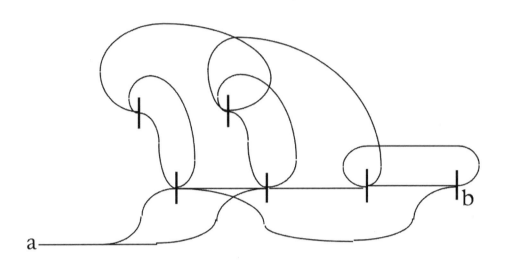

Spencer-Brown's Minimal Modulator

VI. The Balance Game

The balance game is a way to watch the time-behavior of inverter circuits. Here are the rules:

1. Draw a large diagram of the network.
2. Place black and white counters on the inputs and just to the right of each inverter mark. If your choice of counters gives an unbalanced state, follow rule 3 to search for balance.
3. Without changing any input, find the unbalanced marks. Choose one. Balance it by changing its counter. Now there may be a new set of unbalanced marks. If so, choose again and keep doing this until the net is balanced (or until it is clear that no balance can be obtained).
4. Suppose the net is balanced. Then change an input. Now return to step 3 and play out the new transition to balance. Keep a record of the balanced states and the transitional behavior.

This procedure provides a good way to see how the modulators of Spencer-Brown's Chapter 11 work. As an example, we illustrate one transition in the first Spencer-Brown modulator:

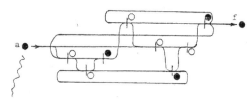

This is balanced. Let a become unmarked.

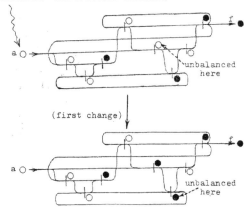

(first change)

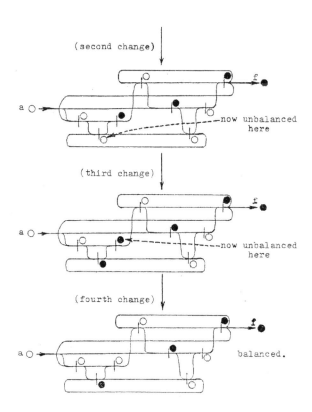

(second change)

(third change)

(fourth change)

balanced.

The network moves to a new balanced state when the input a is changed. These transitions are very easy to work through, using counters (say Go stones) on a circuit diagram. They illustrate the delicate communications relationships that give the network its behavior.

Note how time enters this process. If the marks have different reaction times, this does not affect the behaviors as long as the perturbation of the input a is not too short. Thus in a certain sense, the behavior is timeless, although any particular realization of will have the imperfections of confusion in reaction to overly rapid change or perturbation.

Balance and change: In balance there is fitting together hand-in-glove the sound of one hand clapping. Push. Perturb. Then comes the rush of self-correction, from the swinging of a pendulum or the internal echoes of a Chinese gong, to the clouds of passing thoughts. This reverberation may return to a quiescent state, or there may be a continued sounding, a presence of form.

VII. Imaginary Value

The modulator we discussed in the last section is depicted in chapter 11 of *Laws of Form*. The waveforms in this diagram depict the states at the markers at successive times of stable states in the network. The transitions indicated show the final movement form one stable state to another, and do not indicate the internal structure of the transitions as we have described above. Spencer-Brown has a qualitative description of certain properties of transitions that he calls *imaginary values*. These are places in the circuit where a value lasts just long enough to influence a transition and not any longer! We have illustrated earlier such an imaginary value in the circuit:

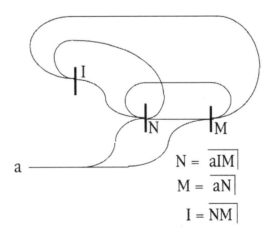

Here is the imaginary value at I that influences the transition of the memory. If a is marked, then N and M are both unmarked. And I is marked. When I becomes unmarked, I holds N to be unmarked and so the transition to the marked state occurs first for M. This results in I immediately becoming unmarked and then N is held unmarked by the value of M. Thus I stays marked just long enough to influence the circuit transition.

Here is the first modulator in chapter 11 of *Laws of Form*.

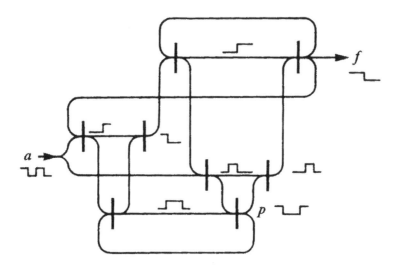

G. Spencer-Brown (1969, p. 66) says: "By using imaginary components of some wave structures, it is possible to obtain the wave structure at p [in the modulator above] with only six markers. This is illustrated in the following equation."

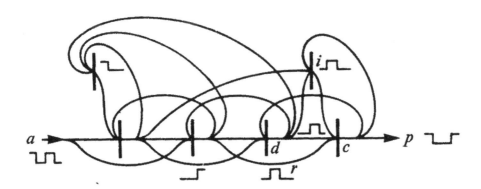

"Here, although the real wave structure at *i* is identical with that at *r*, the imaginary component at *i* ensures that the memory in markers *c* and *d* is properly set. Similar considerations apply to other memories in the expression" (Spencer-Brown, p. 67).

I encourage the reader to play the balance game on this final modulator of Spencer-Brown and to see that indeed there are imaginary value transitions occurring in the network exactly analogous to our example above with a memory that is influenced to go to one state rather than another by the extra observing mark in the circuit.

It should be clear to the reader that this subject of imaginary values has an unsuspected depth. In the Preface to the first edition of *Laws of Form* Spencer-Brown writes:

> Apart from the standard university logic problems, which the calculus published in this text renders so easy that we need not trouble ourselves further with them, perhaps the most significant thing, from the mathematical angle, that it enables us to do is to use complex values in mathematical logic. They are the analogs, in ordinary algebra, to complex numbers $a + b\sqrt{-1}$. My brother and I had been using their Boolean counterparts in practical engineering for several years before realizing what they were. Of course, being what they are, they work perfectly well, ...
>
> Suppose we assume that a statement falls into one of three categories, true, false, or meaningless, and that a meaningful statement that is not true must be false, and one that is not false must be true. The statement under consideration ['This statement is false.'] does not appear to be meaningless ... so it must be true or false. If it is true, it must be, as it says, false. But if it is false, since this is what it says, it must be true.
>
> "It has not been hitherto noticed that we have an equally vicious paradox in ordinary equation theory, because we have carefully guarded ourselves against expressing it this way. Let us now do so."
>
> "We will make assumptions analogous to those above. We assume that a number can be either positive, negative, or zero. We assume further that a nonzero number that is not positive must be negative, and one that is not negative must be positive. We now consider the equation $x^2 + 1 = 0$. Transposing, we have $x^2 = -1$ and dividing both sides by x gives $x = -1/x$."
>
> "We can see that this (like the analogous statement in logic) is self-referential: the root-value of x that we seek must be put back into the expression from which we seek it."
>
> "Mere inspection shows us that x must be a form of unity, or the equation would not balance numerically. We have assumed only two forms of unity, +1 and -1, so we may now try then each in turn. Set x = +1. This gives +1 = -1/+1 = -1 which is clearly paradoxical. So we set x = -1. This time we have -1=-1/-1 = +1 and it is equally paradoxical."
>
> "Of course, as everybody knows, the paradox in this case is resolved by introducing a fourth class of number, called *imaginary*, so that we can say the roots of the equation above are +i and -1, where i is a new kind of unity that consists of a square root of minus one."
>
> "What we do in Chapter 11 is extend the concept to Boolean algebras, which means that a valid argument may contain not just three classes of statement, but four: true, false, meaningless and imaginary. The implications of this, in the fields of logic, philosophy, mathematics, and even physics are profound." (Spencer-Brown, 1969, p. xi)

I have quoted this passage from the Preface to *Laws of Form*, so that you, the reader of this essay, can examine it in the light of our common experience (by now) of the essentials of chapter 11. It may not be quite obvious how Spencer-Brown's remarks about imaginary values in the above quote and his identification of imaginary values in the circuits are related. This we must discuss.

First of all consider the memory as a form of equation $f = \overline{f a \overline{b}}$. As a form of equation, the memory must be put into itself. As a circuit we take for granted that it has that circularity and that this structure gives it the two basic stable states that we have discussed. But as an equation it is self-referential, and subject to the worries of possible paradox. Furthermore, the simpler equation $J = \overline{J}$ as a circuit is an inverter put into itself and its behavior is to oscillate in time between the marked and unmarked states. J asserts that it is not J, and so is an exact analog of the liar paradox:

"This statement is false." And J is an exact analog of the equation for the square root of negative unity x = -1/x, as given by Spencer-Brown above.

Now consider the temporal behavior of $J = \overline{J}$. Its value at any given time is just sufficient to result in the transition to the opposite of that value. This is a circuit with a pure imaginary value. It has no stable state only transitional states that lead to next transitional states. J embodies pure time. Paradox is the generator of time. Time resolves paradox. The re-entering mark gives us a touchstone for examining imaginary values. In the next section we will embark on a description of what comes about by just exploring the imaginary values related to this single circuit.

The subtle question of the uses of imaginary values in the design of modulators such as the ones we have been considering is an open field of research. There is, at the present time, no systematic theory for the design of such asynchronous modulator circuits. Boolean algebra is helpful in local parts of the design of such a circuit, but has to be abandoned to a careful checking of behavior to understand global issues. There are many cybernetic concerns that are mirrored in this design process. Just as we have described in the first part of this essay, one may begin with ideas for how a circuit should behave, and then discover that it will not complete its behavior automatically without the intervention of an outside observer. But then it is often possible to insert marks into the circuit that perform the distinctions of that outside observer. In this way we build a circuit that observes itself, possibly using imaginary values, and behaves as a deterministic asynchronous design.

In the next section we shall embark on a journey to analyze imaginary values beginning with the re-entering mark $J = \overline{J}$. But, having quoted Spencer-Brown on this subject, we should address the possibility that imaginary values can be used to prove mathematical results that would not be reached by the usual Boolean values. He says (see above) "a valid argument may contain not just three classes of statement, but four: true, false, meaningless and imaginary." Can we prove a theorem using imaginary values? What would this mean?

Some examples from mathematics may help. But in order to use the mathematical framework I am going to take a slightly different beginning point for imaginary values. I will talk about values or constructions that are seen to not belong to the original domain where one started to work. This is in line with Spencer-Brown's analogy that $x^2 = -1$ does not have a solution in the domain of unit values $\{-1, +1\}$, and so we make up a new imaginary value i that does satisfy $i^2 = -1$. We then find out the consequences of this value and that we can prove theorems using it. There is also a subtle time-shift associated with i as we shall see in the next section of the paper. It seems to be the case that time enters in special ways when one is involved in the construction of a new domain that goes beyond the domain where one has become accustomed to live. In the case of the modulators, we needed to create a new observer for the given modulator and to insert that observer into the modulator itself. The design worked in the temporality of that insertion.

Lets go back to i for a moment. We have complex numbers of the form z = a + bi where a and b are real numbers and $i^2 = -1$. We define the conjugate of z to be z* = a −

bi. It is easy to check that $zz^* = a^2 + b^2$ and that $zw = wz$ and $(zw)^* = z^*w^*$ for any complex numbers z and w. Suppose $z = a + bi$ and $w = c + di$. Then

$$(a^2 + b^2)(c^2 + d^2) = zz^* ww^* = zw\, z^*w^* = zw\, (zw)^*.$$

But $zw\, (zw)^*$ is a sum of two squares (the details of which can be found by multiplying out zw). And so we have proved the

Theorem. The product of a sum of two squares is a sum of two squares.
Our proof was a proof by using imaginary values. We used the assumed existence and properties of the square root of minus one to prove a theorem about real numbers and in fact a theorem about integers, since we can take a,b,c,d to all be integers in this work.

Of course we can do the work suggested by our imaginary argument and form $zw = (a + bi)(c + di) = ac - bd + (ad + bc)i$ and so we see that we have proved

$$(a^2 + b^2)(c^2 + d^2) = (ac - bd)^2 + (ad + bc)^2.$$

For example,
$$(3^2 + 4^2)(5^2 + 7^2) = 13^2 + 41^2.$$

Of course any ordinary mathematician can now verify directly this algebraic formula showing that the product of two sums of two squares is itself a sum of two squares. Our reasoning by imaginary values gave the result without the need for direct calculation.

And there is more. Suppose $a = c$ and $b = d$. Then we have proved that

$$(a^2 + b^2)^2 = (a^2 - b^2)^2 + (2ab)^2.$$

This is the well-known formula for producing Pythagorean triples (C, A, B) such that $C^2 = A^2 + B^2$. When a and b are integers, then so are $C = a^2 + b^2$, $A = a^2 - b^2$ and $B = 2ab$. For example,
$$(3^2 + 2^2)^2 = (3^2 - 2^2)^2 + (2 \times 3 \times 2)^2.$$

Hence
$$13^2 = 5^2 + 12^2.$$

We have produced a number of stunning elementary results in number theory via reasoning with imaginary values, where these values are the system of complex numbers using i with $i^2 = -1$.

In the present day this is not usually regarded as a new form of reasoning. We have become used to the situation that one form of mathematics may become a structure through which truths in another form of mathematics can be seen. We have convinced ourselves that the complex numbers are a consistent form of mathematics and that it is legitimate to use them in proving theorems in realms that are not directly related to the complex numbers. Furthermore, most people who codify logic (logicians, if you will) do not regard the complex numbers as a form of reasoning. Here, I am suggesting that each form of mathematics is a form of reasoning! And sometimes a mathematical form can be used to prove a result in a domain where its applicability was not expected! In that sense one form of mathematics can act as an imaginary value for another form. This is what we have seen here where the imaginary value i informs the theory of numbers.

Here is another example. Consider the well-known proof that the square root of 2 is irrational. This proof goes as follows:

> Suppose S is a number such that $S^2 = 2$ and that $S = P/Q$ where P/Q is a positive rational number in reduced form so that the integers P and Q have no common factor. Then $2 = P^2/Q^2$ and hence $2P^2 = Q^2$. From this it follows that 2 divides Q^2 and hence that 2 divides Q (since the square of an odd number is odd). But if 2 divides Q, then $Q = 2R$ for some integer R, and so $2P^2 = 4R^2$ and therefore $P^2 = 2R^2$. This implies that 2 divides P. We have shown that both P and Q are divisible by 2. This is a contradiction, since we assumed at the beginning that they had no common factor. Therefore we have proved by contradiction that S cannot be rational.

If all numbers were assumed to be rational, this proof would have an entirely different character. It would be a paradox!

Given that $S^2 = 2$, and surely S is rational, then we obtain a contradiction! This is how it appeared to the Greek mathematicians who discovered this argument. To them the square root of two existed because, by the theorem of Pythagoras, it was the length of the diagonal of a square of side one. Thus it was given that there is a square root of two. And for them, all numbers should be rational numbers. But the argument shows that the square root of two is irrational. This was a terrible contradiction. And the square root of two had to become a new sort of number. It was called a surd (absurd) or irrational. There was at that time no theory for these new numbers. They were imaginary.

And time does enter in this story! For those same Greeks discovered that they could find very beautiful approximations to the square root of two by using a recursive process. The final answer was $T = 1+\sqrt{2} = 2 + 1/(2+ 1/...)$. This infinite continued fraction reenters its own computational space as

$$T = 2 + 1/T$$

and so must satisfy the quadratic equation $T^2 = 2T + 1$ and you can easily see that this is indeed satisfied by $1+\sqrt{2}$.

On the other hand, if $T = 2 + 1/T$, then

$$T = 2 + \frac{1}{T} = 2 + \cfrac{1}{2+\cfrac{1}{T}} = 2 + \cfrac{1}{2+\cfrac{1}{2+\cfrac{1}{T}}} = \ldots$$

and we can formally write the infinite continued fraction

$$T = 2 + \cfrac{1}{2+\cfrac{1}{2+\cfrac{1}{2+\ldots}}}$$

This imaginary T sits inside itself in the infinite construction and so satisfies $T = 2 + 1/T$. But in fact, the finite approximations to the infinite continued fraction do produce numbers that get closer and closer to $1+\sqrt{2}$. And we declare

$$1+\sqrt{2} = 2 + \cfrac{1}{2+\cfrac{1}{2+\cfrac{1}{2+\ldots}}}$$

What we have done to produce a square root of two (1 plus that number) by using continued fractions is an exercise in the mathematical imagination. We have supposed that there was such a number and then found out what it must be. But it is more than that. After all, according to our continued fraction, the series

$$2, 2 + \tfrac{1}{2} = 5/2, 2 + 2/5 = 12/5, 2 + 5/12 = 29/12, 2 + 12/29, \ldots$$
$$2, 5/2, 12/5, 29/12, 70/29, \ldots$$

should become better and better approximations to $1+\sqrt{2}$, and they do! So the series of numbers $1, 5, 12, 29, 70, \ldots$ is a key to this imaginary entity the square root of two. And these numbers have the simple pattern that any one of them is equal to twice its predecessor plus the one before that. For example $29 = 2 \times 12 + 5$.

By assuming the existence of the square root of two, we have brought it into existence as an infinite continued fraction and we have found an infinity of rational approximations to it. This entire mathematical process of discovery is seen as a form of reasoning using imaginary values. Time enters into the process by the very steps needed to examine the structure of the reentry of the number into its own form.

VIII. Time and Imaginary Value – Starting with J = <J>.

Now lets start with J = <J> again. We have already understood that we can think of this as a series of states in time that alternate between being marked and being unmarked. Alternatively, as with the continued fraction for $1+\sqrt{2}$ in the last section, we can examine the identities that J must satisfy, but substituting J into its own indicational space. We obtain

$$J = <J> = <<J>> = <<<J>>> = <<<<J>>>> = <<<<<J>>>>> = \ldots$$

This is also a temporal sequence, a building up of spatial forms. To someone with the idea of a never-ending, infinite, nest of marks it suggests that we could take J = <<<<<…>>>>>. Then adding one external mark to the infinite nest will not change the infinity and so we would have J = <J> as a consequence of the infinite construction!

Should J be a logical value, it can be neither True nor False since it declares itself to be the opposite of any value that is assigned to it. Thus J is an exemplar of an imaginary logical value. Can we enfold this value in the Boolean context? And yet and yet, when we go to infinity something new occurs.

The essence of memory is the circuit M = <<M>>. There is no paradox here. If M is marked then <<M>> is also marked. If M is unmarked, then so is <<M>>. And yet we can perform the ritual of recursive substitution once again

$$M = <<M>> = <<<<M>>>> = <<<<<<M>>>>>> = \ldots$$

and, taking the process to infinity we have a limit M of the form

$$M = <<<<<<<<<<\ldots>>>>>>>>>>.$$

It would seem that in the limit M and J arrive at the same form. Any finite version of M has an even number of marks, while the finite versions of J can have an even or an odd number of marks. When we write J = <J> = <<J>> we cannot take J = <<J>> as the defining equation for J. This equation states only that J = J and says nothing more. The infinite form has two states that do not oscillate, just as in the case of our finite memory. I will indicate them here by decorating the form.

$$<<<<<<<<\ldots>1>0>1>0>1>0>1>0$$

$$<<<<<<<<\ldots>0>1>0>1>0>1>0>1$$

Once labeled, the infinite form has two states, one "even" and one "odd". The basic memory is implicated in the limit structure of the imaginary value that is the re-entering mark. The alternating process of 0 and 1 has become the infinite labeling of the infinite form.

We now understand that the memories that build the modulators are themselves imaginaries and that the whole construction, leading to the underpinnings of all the computers in the world is imaginary.

IX. Waveform Arithmetics and the Square Root of Negation

Start again with $J = <J>$ and regard it as generating the series …0101010101…. Here 0 and 1 are the two Boolean values. Just as we ended the last section, there are really two series, one that starts with 0 and one that starts with 1:

$$I = 0101010101010…$$

$$J = 10101010101010…$$

At any given time the two series have opposite values so we could say that $IJ = 0$, multiplying them together term by term. In comparing them, we do not really need a starting point.

$$I = … 0101010101010…$$

$$J = … 1010101010101…$$

It is the relative condition of the two series that makes the difference. So you can say that $<I> = I$ and $<J> = J$ when looking at them individually, but when you put them together you see the shift (analogous to a phase shift for waves) from one to the other. This leads to the construction of a finite language that can distinguish these states. We can write an *ordered pair* $I = (0,1)$ and $J = (1,0)$ and more generally make ordered pairs (a,b) and (b,a) for a series …abababababab… that can be viewed in these two phase shifted ways. We can define combinations of the processes by the equations

$$(a,b)(c,d) = (ac,bd) \text{ and } (a,b)+(c,d) = (a+c,b+d).$$

Then $IJ = (0,1)(1,0) = (0,0) = 0$ and $I + J = (0, 1)+(1, 0) = (0+1,1+0)= (1,1)$ and this is correct if the underlying arithmetic is Boolean or standard. Note that in Boolean arithmetic $1 + 1 = 1$ and that we take + as "or" and multiplication as "and" in the standard Boolean form.

There is a nice way to have $<I> = I$ and $<J> = J$.
We define $<(a,b)> = (<b>,<a>)$.
Note how this works: $<I> = <(0,1)> = (<1>,<0>) = (0,1) = I$. By using ordered pairs, we have achieved self-reference without going to infinity. The idea is very simple. We understood that if we just changed the waveform by flipping it at every individual time, then I would change to J and J would change to I. But if we also shifted each waveform by one phase, then J would change to I and I would change to

J. So we do both operations. We cross each element of the waveform and we shift the waveform by one time step. This takes us right back to where we started and self –reference is the result. The resulting waveform arithmetic and its algebra is very interesting. The reader can examine (Kauffman & Varela, 1980).

Implicitly in making this construction we used a new operator that I shall denote by (a, b)* = (b, a). This is the time-shift operator. Along with the time-shift, we define [(a, b)] = (<a>,<b>), the *standard inversion* of (a,b). The standard inversion shifts the waveforms and so [I] = J and [J] = I.

Then we have that <(a,b)> = (<b>,<a>) = [(b,a)] = [(a,b)*] and so we can write

$$<X> = [X*]$$

so that

$$<I> = [I*] = [J] = I \text{ and } <J> = [J*] = [I] = J.$$

There is a rich possibility in the interactions of the waveforms that emanate from self-reference and imaginary value. If we define

$$\#(a,b) = (<b>,a)$$

then we obtain a *square root of negation* in precise analogy with the way that Sir William Rowan Hamilton defined the square root of minus one in terms of numerical ordered pairs. Note how this works:

$$\#\#(a,b) = \#(<b>,a) = (<a>,<b>) = [(a,b)].$$

Thus

$$\#\#X = [X]$$

for any ordered pair X.

The numerical analogy is i(a,b) = (-b,a) and then ii(a,b) = i(-b,a) = (-a,-b) = -(a,b) so that ii = -1. This was how Hamilton defined i using ordered pairs of numbers.

See the following papers for an early version of this idea: Kauffman (1985a, 1987a, 1987b); and see the paper Kauffman (2017) for our work with Art Collings (2017) making the Boolean version into both an ordered pairs calculus and a containment calculus close to the spirit of Spencer-Brown's laws of form.

To make this formalism for the square root of negation work in a containment calculus we can use the following notation, with a right mark operating as standard negation and the left mark operating as a square root of standard negation.

$$\overline{(a,b)|} = (\overline{a}|,\overline{b}|)$$

$$\overline{|(a,b)} = (\overline{b}|,a)$$

Then we see that

$$\overline{|\overline{|(a,b)}} = \overline{|(\overline{b}|,a)} = (\overline{a}|,\overline{b}|) = \overline{(a,b)|}$$

and so we have the operator equation $\lceil\lceil = \rceil$, and the left mark $\lceil$ is a square root of the right mark $\rceil$.

In this way we have a new arithmetic with four values: the unmarked state, the left mark $\lceil$, the right mark $\rceil$ and the triple left mark $\lceil\lceil\lceil = \rceil\rceil = \rceil$. In this arithmetic $\rceil_a = \rceil$ for any a, and $\lceil\lceil\lceil = \lceil\rceil\rceil = \rceil$. This is the Collings arithmetic. The reader will see that it is compatible with our ordered pair interpretation via the identifications:

$$(,) = \text{unmarked state.}$$

$$\lceil = \overline{(,)} = (\rceil,)$$

$$\lceil\lceil = (\rceil,\rceil) = \rceil$$

$$\lceil\lceil\lceil = (\overline{\rceil},\rceil) = (,\rceil)$$

The key point here is that a simple containment calculus extending the laws of form calculus has a strong connection with the fundamental structure of the complex numbers and the square root of negative unity.

Remark on Zero: Start in a Boolean realm with 0 and 1 and ~0=1 and ~1=0 (we use the logical negation sign here). Then it is a big step to extend the domain and allow self referential entities like J with ~J = J. In ordinary arithmetic of real numbers the minus sign as operator is the analogue of negation with –(1) = -1 and –(-1) = 1. Thus -1 and 1 are the numerical analogues of 0 and 1. But in numerical arithmetic we have a different role for 0 and indeed -0 = 0. Thus in the numerical context zero is a self-referential value and we do not regard it as paradoxical. The extremes of +1 and -1 are mediated by 0 and zero is unchanged under negation. Self reference in the ordinary mathematical domain is commonplace. Self reference in Boolean domains has been regarded with suspicion. We suspect that such suspicion is misplaced.

X. The Series ...+,-,+,-,+,-, ... Temporal Calculus, Complex Numbers and Clifford Algebra

Now we are going to shift to numerical rather than Boolean sequences. The lead for this is Spencer-Brown's analogy that writes the equation $i^2 = -1$ as

$$i = -1/i$$

so that we have the mapping $T(x) = -1/x$ from the real numbers to the real numbers. There is no real number that satisfies the equation $x = T(x)$. If $x = 1$ then $T(x) = -1$. If $x = -1$, then $T(x) = 1$. We are confronted with the series

$$...-1 +1 -1 +1 -1 +1 -1 +1 -1 +1 ...$$

and once again we shall take ordered pairs so that this series has two views

$$\varepsilon = (-1, 1) \text{ and } \varepsilon^* = (1,-1) = -(-1,1) = -\varepsilon.$$

Here we allow multiplication of the ordered pairs by constants so that $k(a,b) = (ka,kb)$. We want to understand that this series should represent the square root of minus one. In fact we can transpose the ideas we used in designing waveform arithmetics to do the job. I will introduce a new notation $\{a,b\}$ as a time sensitive ordered pair.

If you multiply $\{a,b\}$ by another time sensitive entity $\{c,d\}$, then $\{c,d\}$ will be shifted by one time step in the interaction. Here are the rules, where x and y are ordinary ordered pairs.

1. $\{x\}\{y\} = xy^*$
2. $x\{y\} = \{xy\}$
3. $\{x\}y = \{xy^*\}$

I have indicated the temporal rules. Otherwise we have $\{x+y\} = \{x\}+\{y\}$ and we write $\{(a,b)\} = \{a,b\}$ since the braces are just a way to indicate that an ordered pair is temporal.

The rules are consistent with the idea that we operate from left to right and that for time sensitive entities like $\{x\}$, the clock must tick when we use them. Multiplying $\{x\}$ and $\{y\}$ temporally, we have to take y at the next time, and this is y^*. Thus

$$\{a,b\}\{c,d\} = (a,b)(c,d)^* = (ad,bc).$$

Lets concentrate on the first rule $\{x\}\{y\} = xy^*$. Try it out for $i = \{\varepsilon\}$. Then

$$ii = \{\varepsilon\}\{\varepsilon\} = \varepsilon\varepsilon^* = (-1,1)(1,-1) = (-1,-1) = -1.$$

This shows that we can define $i = \{\varepsilon\}$ and obtain the square root of minus one! Defining i in this way makes i a time sensitive entity. When i interacts with itself, it shifts one of the copies by one time step and so produces plus and minus together at each meeting point. In this way we have interpreted the sequence

$$\ldots -1 +1 -1 +1 -1 +1 \ldots$$

as the square root of minus one. The imaginary value makes just the right step in time at the right time.

The system we have created consists of numbers of the form $x + \{y\}$ where x and y are standard ordered pairs. We have in fact created much more than the complex numbers by these temporal interaction rules. But let us look at some properties of the system.

Let $\eta = \{1\} = \{(1,1)\} = \{1,1\}$.
Then $(a,b)\eta = (a,b)\{1\} = \{a,b\}$

but $\eta(b,a) = \{1\}(b,a) = \{1(b,a)^*\} = \{a,b\}$.
Thus $(a,b)\eta = \eta(b,a)$.

This tells us that η acts as a time shifter. If you change the order of multiplication with η, then you get shifted by one time step. Note also that $\eta\,\eta = \{1\}\{1\} = 1$ so this is also in accord with our time shifting. Two time shifts and we are back where we started, since the pattern of the process has period two. In this form we can write $\eta(a,b)\,\eta = (b,a)$ or in other words $\eta x \eta = x^*$.

Note also that since $(a,b)\eta = \{a,b\}$ we can dispense with the curly parentheses and just write $(a,b)\eta$ instead. Then the elements of our system are of the form $x + y\,\eta$ where x and y are ordered pairs.

We can then check that this new multiplication is associative. That is if $A = x + y\,\eta$, $B = z + w\,\eta$ and $C = r + s\,\eta$, then $A(BC) = (AB)C$. I will leave this for you to work out for yourself. Here is the main part. Suppose that $A = \{x\}, B = \{y\}, C = \{z\}$. Then

$$A(BC) = \{x\}(\{y\}\{z\}) = \{x\}yz^* = \{x(yz^*)^*\} = \{x\,y^*z^{**}\} = \{xy^*z\} = xy^*\{z\} = (\{x\}\{y\})\{z\} = (AB)C.$$

The temporal rules conspire to produce an associative algebra. If you are familiar with matrix algebra then you can see that $(a,b) + (c,d)\,\eta$ behaves exactly like a 2 x 2 matrix

$$(a,b) + (c,d)\,\eta = \begin{pmatrix} a & c \\ d & b \end{pmatrix}.$$

I am particularly interested in the following elements of this temporal algebra:

$$1,\ \varepsilon,\ \eta,\ \varepsilon\,\eta.$$

We have seen that $\varepsilon\,\eta = \eta\,\varepsilon^* = \eta(-\varepsilon) = -\eta\,\varepsilon$. Thus $\varepsilon\,\eta + \eta\,\varepsilon = 0$. Note that $\varepsilon\,\varepsilon = \eta\,\eta = 1$ and that $(\varepsilon\,\eta)^2 = -1$. The two fundamental elements ε and η anticommute. We have constructed more than just the complex numbers.

The algebra generated by $1,\ \varepsilon,\ \eta,\ \varepsilon\,\eta$ is called a Clifford algebra (Kauffman, 2016a) and in this case is often called the *split quaternions*. By allowing an extra square root of minus one that commutes with the generators of the Clifford algebra, we can easily construct the quaternions. To see this, suppose that $\iota\,\iota = -1$ and that ι commutes with e and η. Let

$$I = \iota\,\varepsilon,\ J = \iota\,\eta,\ K = \eta\,\varepsilon.$$

You can easily check that $I^2 = J^2 = K^2 = -1$. For example,

$$I^2 = \iota\,\varepsilon\,\iota\,\varepsilon = \iota\,\iota\,\varepsilon\,\varepsilon = -1.$$

And we also have

$$IJK = \iota\,\varepsilon\,\iota\,\eta\,\eta\,\varepsilon = \iota\,\iota\,\varepsilon\,\eta\,\eta\,\varepsilon = \iota\,\iota\,\varepsilon\,\varepsilon = \iota\,\iota = -1.$$

Thus
$$I^2 = J^2 = K^2 = IJK = -1$$

and this defines the quaternions.

We have shown that basic Clifford algebra and the quaternions arise naturally from the imaginarity of the recursion associated with the square root of minus one. All of this as a consequence of the dynamics the reentering mark.

XI. The Series ...010101010101... and the Fermion Algebra

Let's return now to the series alternating between 0 and 1 with numerical values.

$$...01010101010101...$$

We apply the same analysis as in the last section. Take p = (1,0) and q = (0,1) as the two views of the sequence. Note that

$$pq = (1,0)(0,1) = (0,0) = 0.$$
$$p + q = (1,0) + (0,1) = (1,1) = 1.$$

We have
$$U = p\eta \text{ and } U^\dagger = q\eta$$

as the two basic temporal elements for the algebra of this series. Note that
$$p\eta = (1,0)\eta = \eta(0,1) = \eta q,$$

$$q\eta = \eta p.$$

Hence $\eta p \eta = q$ and $\eta p \eta = q$, making p and q time shifts of each other. Then

$$U^2 = p\eta p\eta = pq = 0,$$

and
$$(U^\dagger)^2 = q\eta q\eta = qp = 0.$$

Furthermore
$$UU^\dagger = p\eta q\eta = pp\eta\eta = pp = p,$$

$$U^\dagger U = q\eta p\eta = qq = q.$$

Hence
$$UU^\dagger + U^\dagger U = p + q = 1.$$

Thus we have shown that

$$U^2 = (U^\dagger)^2 = 0 \text{ and } UU^\dagger + U^\dagger U = 1.$$

These are the fundamental equations for the creation and annihilation operators for a Fermionic particle in quantum mechanics (Kauffman, 2016a). More than one Fermi particle cannot occupy the same space with another. This is the moral of the equations where the square of an operator is 0. Fermi particles can (irrespective of order) be created in pairs from the vacuum. This is the moral of the second equation where pairs $UU^\dagger$ or $U^\dagger U$ can appear from 1 (the vacuum).

Fermions embody very special distinctions and it is very beautiful that we find them emerging from the process algebra for a single distinction. In the next section we shall see how indeed how Fermions emerge from the Clifford algebra associated with the numerical plus minus sequence. Here is the zero-one amazement and the plus-minus amazement in dialogue together.

For now, look more closely at the Fermionic algebra. We have

$$UU^\dagger U = (1 - U^\dagger U)U = U - U^\dagger UU = U$$

since $U^2 = 0$, and so $UU^\dagger = p$ we have

and
$$UU^\dagger U = U$$
$$U^\dagger UU^\dagger = U^\dagger.$$

These equations tell us that concatenations of these operators will only produce the possible values
$$0, 1, U, U^\dagger, UU^\dagger, U^\dagger U.$$

We now have, along with the individual distinction of 0 and 1, the new distinction between $UU^\dagger = p$ and $U^\dagger U = q$ and how the operators move these around.

In particular look at the diagram below the arrows mean that you multiply by the label of the arrow on the left.

$$0 \xleftarrow{U} U^\dagger U \underset{U}{\overset{U^\dagger}{\rightleftarrows}} U^\dagger \xrightarrow{U^\dagger} 0$$

Let A denote $U^\dagger U$ and B denote $U^\dagger$. Then we see

$$0 \xleftarrow{U} A \underset{U}{\overset{U^\dagger}{\rightleftarrows}} B \xrightarrow{U^\dagger} 0$$

We see two distinct entities or states A and B (the two spin states of the Fermion particle) and the operators U and $U^\dagger$ that exchange these states. If you apply U twice you get zero. If you apply $U^\dagger$ twice you get zero. But if you apply the operators alternately, then you bounce back and forth between the two spin states. Each operator U or $U^\dagger$ satisfies the law of crossing: It operates on itself to annihilate itself. But the two operators together form the acts of crossing between two sides of the distinction between A and B. We see that the Fermionic operators make the one of the first departures from the form in that we now distinguish the direction of crossing, either from A to B or from B to A. These elementary distinctions and the Fermion algebra all arise naturally from the recursive process of reentry for a single distinction.

A Fermionic Containment Calculus
We can make a Fermionic containment calculus by first defining two new operators on laws of form pairs. We define

$$\lfloor(a,b) = (b, \rceil)\text{, the left push,}$$

and

$$(a,b)\rfloor = (\lceil, a)\text{, the right push.}$$

Note that for any (a,b), two pushes of the same type carry it to the marked state.

$$\lfloor\lfloor(a,b) = \rceil$$

$$(a,b)\rfloor\rfloor = \rceil$$

The left and right push operators do not commute,

$$\lfloor(a,b)\rfloor = (a, \rceil)$$

$$\lfloor(a,b)\rfloor = (\lceil, b)$$

however,

$$\lfloor(a,b)\rfloor + \lfloor(a,b) = (a,b).$$

Where $x + y = \overline{\overline{x}|\overline{y}|}$, we have

$$\rfloor\rfloor = \lfloor\lfloor\equiv\rceil$$

$$\lfloor\rfloor + \lfloor\rfloor =$$

and $\rceil\rfloor = \rceil = \square$ while $\square = \rfloor$ and $\rceil\rfloor = \lfloor$.

Thus we have a non-commutative Fermionic calculus of containment operators. There is an algebra and an arithmetic to be explored, but we shall not do that in this paper. By taking $I = (\lceil,)$ and $J = (, \rceil)$, we have the schematic containment calculus version of our Fermion diagram above.

$$\rceil \xleftarrow{\lfloor} I \xrightleftharpoons[\rfloor]{\lfloor} J \xrightarrow{\rfloor} \rceil$$

The objects I and J are exchanged by the left and right pushes, but when a push is applied twice one ends up at the marked state, as indicated at the left and the right of the diagram. This diagram with arrows could be called the *Fermion category*, a small structure with arrows and compositions of arrows that embodies the essence of the Fermion operator algebra. See MacLane (1986) for a discussion of category theory.

XII. The Fermion Mystery

I am going to show you a very strange property of the Clifford algebra that we have discussed in previous sections. Suppose that we write a general element of this algebra in the form $U = \varepsilon A + \eta B + \varepsilon \eta C$ where A, B and C are real numbers. Now compute the square of this element.

$$U^2 = (\varepsilon A + \eta B + \varepsilon \eta C)^2 =$$

$$(\varepsilon A)^2 + (\eta B)^2 + (\varepsilon \eta C)^2 + (\varepsilon \eta + \eta \varepsilon)AB + (\varepsilon \varepsilon \eta + \varepsilon \eta \varepsilon)AC + (\eta \varepsilon \eta + \varepsilon \eta \eta)BC.$$

And you will find that just as $(\varepsilon \eta + \eta \varepsilon) = 0$, all the other cross terms vanish. Since $\varepsilon^2 = \eta^2 = 1$ and $(\varepsilon \eta)^2 = -1$, we get

$$U^2 = A^2 + B^2 - C^2.$$

We have proved the following result:

> Theorem. If $U = \varepsilon A + \eta B + \varepsilon \eta C$ then $U^2 = A^2 + B^2 - C^2$.
> Hence $U^2 = 0$ if and only if $C^2 = A^2 + B^2$.

This suggests that the Pythagorean theorem is involved with our algebra for the imaginary states of the reentering mark. Who ordered that? And what would it mean for U to have square zero? An operator with square zero is called a *nilpotent operator*.

We started with a nilpotent operator in the form of the mark: ⏌ = .

We took a long journey in articulating the properties of the temporal behavior of the reentering mark and arrived at a Clifford algebra where the general element U of that algebra acts upon itself to produce zero, exactly when the parameters of that element U satisfy the Pythagorean theorem!

In a philosophical mode we could reason that if U is the Universe then U can act upon itself to annihilate itself. In the algebraic operator mode of speaking, this is in the form of

$$UU = \text{Nothing}$$

or

⏌ = .

Thus we have found a primordial algebraic expression $U = \varepsilon A + \eta B + \varepsilon \eta C$, where $C^2 = A^2 + B^2$, for the *fundamental vanisher*, the representative of universe that creates from nothing and reverts back to nothing.

$$U^2 = 0 \text{ when } C^2 = A^2 + B^2.$$

In the last section we found operators U and $U^\dagger$, both of square zero and so that $UU^\dagger + U^\dagger U = 1$, corresponding to the algebra of a Fermion. In the present context there is a good choice for $U^\dagger$ (and we shall explain below why this is a good choice). It is

$$U\dagger = \varepsilon A + \eta B - \varepsilon \eta C.$$

You can easily check that $(U^\dagger)^2 = 0$. Note that we have the calculation

$$UU^\dagger + U^\dagger U = (U + U^\dagger)^2 = 4(\varepsilon A + \eta B)^2 = 4(A^2 + B^2) = 4C^2.$$

And so (up to a constant) U and $U^\dagger$ also satisfy the Fermion equations, and they have the variability afforded by solutions to the Pythagorean equation $A^2 + B^2 = C^2$.

We might stop here with this story of the return to the nilpotent mark after a long journey into algebra, but there is a relationship with physics that must be mentioned. First of all, there is the remarkable formula

$$E^2 = (pc)^2 + (mc^2)^2.$$

This is the Einstein formula for the energy of a particle with rest mass m that has a momentum p with respect to a given observer [Kauffman, 2016a]. We usually think of the formula $E = mc^2$ for the relationship of mass and energy, but if the mass is moving past us at momentum p, then there is a Pythagorean relationship of the total energy E and the energy mc^2 of the stationary particle. Thus we have the relativistic formula

$$E = \sqrt{(pc)^2 + (mc^2)^2}$$

making the energy the length of the hypotenuse of a right triangle whose sides are pc and mc^2.

Here, for the record is how to see the Einstein formula: We have $E = m_0 c^2$ where m_0 is the rest mass of the particle. And we have $p = mv^2$ where m is the relativistic mass with

$$m = m_0 / \sqrt{1 - v^2 / c^2}.$$

The formula then follows by squaring this (Pythagorean) mass equation:

$$m^2 = m_0^2 / (1 - v^2/c^2)$$
$$m^2(1 - v^2/c^2) = m_0^2$$
$$m^2(c^2 - v^2) = m_0^2 c^2$$
$$(mc^2)^2 = (mv)^2 c^2 + m_0^2 c^4$$
$$E^2 = p^2 c^2 + m_0^2 c^4$$

What will happen if we combine our fundamental nilpotent U with this Pythagorean energy formula from special relativity? The remarkable answer is that we arrive at fundamental solutions to Dirac's relativistic equation for the electron.

To make the formalism easier, let us take c = 1. We use the convention that the speed of light is equal to 1. Then $E^2 = p^2 + m^2$ where m is the rest mass of the particle. With this we could write $U = p\varepsilon + m\eta + E\varepsilon\eta$ and obtain a physical nilpotent element. It follows at once from our theorem that $U^2 = 0$. We are hoping to find an intuitive association of our Clifford algebra with the concepts of energy, momentum and mass. So we have associated momentum with ε, the representative of the on-going time series of the recursion. And we have associated mass with the time shift operator and energy with the time sensitive operator whose square is minus one.

We shall define the dual nilpotent $U^\dagger = p\varepsilon + m\eta - E\varepsilon\eta$. Here time goes backwards relative to U, and if U creates a particle, then $U^\dagger$ creates the corresponding anti-particle! See the discussion below.

We then find (as above) that

$$UU^\dagger + U^\dagger U = (U + U^\dagger)^2 = 4(p\varepsilon + m\eta)^2 = 4(p^2 + m^2) = 4E^2$$

Thus we have that

$$UU^\dagger + U^\dagger U = 4E^2$$

and

$$U^2 = U^{\dagger 2} = 0.$$

The first equation expresses that a creation of a particle and an antiparticle (in either order) can proceed from pure energy. The second two equations represent the Pauli exclusion principle that forbids the existence of identical particles at the same place. There is just no possibility for the production of two identical Fermi particles from pure energy.

We began this essay with a tip of the hat to the Pauli exclusion principle, and we return to it in this way at the end. For the details about how this Fermion algebra is related to the Dirac equation, we recommend that the reader examine Kauffman (2016a) and the work of Peter Rowlands (2007) who discovered the nilpotent approach to the Dirac equation in terms of quaternion algebras.

Here is how to find the Dirac equation from this place. We are looking for a differential operator D such that given $\varphi = e^{i(px-Et)}$ (a plane wave corresponding to momentum p and energy E), then $D\varphi = U\varphi = (p\varepsilon + m\eta + \varepsilon\eta E)\varphi$.

If we have such an operator, then $D(U\varphi) = U^2 \varphi = 0$. So $U\varphi$ would be a solution to the equation $D\phi = 0$. This will be the Dirac equation!

It is easy therefore to just read off the operator D from this condition and conclude that

$$D = -i\varepsilon \partial/\partial x + \eta m + i\varepsilon\eta \partial/\partial t$$

Thus we have constructed a Dirac operator. We define $\hat{p} = -i\partial/\partial x$ and $\hat{E} = i\partial/\partial t$ as the (respectively) momentum and energy operators. These operators extract momentum and energy from the plane wave via

$$\hat{p}\varphi = p\varphi \text{ and } \hat{E}\varphi = E\varphi.$$

Then

$$D = \varepsilon\hat{p} + \eta m + \varepsilon\eta\hat{E}$$

and this it follows that

$$D = \varepsilon\eta(\eta\hat{p} - \varepsilon m + \hat{E}).$$

This means that $D\phi = 0$ is equivalent to saying that

$$\hat{E}\phi = (-\eta\hat{p} + \varepsilon m)\phi.$$

This is the form of the original Dirac equation. Dirac devised his equation so that it would correspond to the basic fact that we have $E^2 = p^2 + m^2$. He needed an operator form of the square root $E = \sqrt{p^2 + m^2}$. Dirac reasoned that if $\alpha^2 = \beta^2 = 1$ and $\alpha\beta + \beta\alpha = 0$, then $(\alpha p + \beta m)^2 = p^2 + m^2 = E^2$ and so he could identify the energy operator with $\alpha\hat{p} + \beta m$.

So in our notation, if we have $\hat{E} = (-\eta\hat{p} + \varepsilon m)$ then $\hat{E}^2 = (-\eta\hat{p} + \varepsilon m)^2 = \hat{p}^2 + m^2$ and we have exactly what is desired to correspond to the Einstein relation. In this way Dirac devised his equation.

We have seen how we can back-engineer Dirac's equation from the requirement that

$$U\varphi = (p\varepsilon + m\eta + \varepsilon\eta E)e^{i(px-Et)}.$$

should be a solution to the Dirac equation! Now you can see why we took

$$U^\dagger = p\varepsilon + m\eta - \varepsilon\eta E,$$

for this corresponds exactly to reversing time in the plane wave and so corresponds creating an anti-particle. The final expression of this Fermionic solution to the Dirac equation $U\varphi = (p\varepsilon + m\eta + \varepsilon\eta E)e^{i(px-Et)}$ is quite remarkable, combining the algebraic form of U and the continuously varying exponential expression in space, time, momentum and energy.

Discussion

We have reached the end of this essay. We began with a fable about the interactions of self referential distinguishers that could wrap themselves around each other and would often disappear due to the enjoinders of the Pauli exclusion principle in a simplified version of the Big Bang. We then got down to business and showed how the little circular distinguishers could pair up and form memories, and then from there they could gang up and form modulators that would reduce the frequency of an oscillation. These modulators were asynchronous, working independently of any particular time delays that might happen in later universes where time had appeared. At this early date there was no hope to produce synchronicity. Why multiplicity itself was just beginning in that era. Before that, there was but one distinction. And before that... nothing. The discipline of designing correct asynchronous modulators occupied us for some time and we developed various approaches to it, not the least of which was the method of modeling that used paper diagrams for the circuitry and Go stones to mark the local states. In the course of this modulator discipline we discovered how imaginary values often make the smallest modulators operate correctly. These imaginary values are states of the circuit that last just long enough to influence a transition, and then disappear to the stable real value of the circuit at that point. Imaginary values occur at the point where the circuit, due to perspicuous design, observes itself to influence its own behavior. We then decided to explore imaginary values and look at what at first appeared to be the narrow road of looking just at the flickering transitions of the reentering mark, a simplest circuit that oscillates always because it feeds back to itself the opposite of its present value. Such a circuit is a pure imaginary. We then found that there are extraordinary arithmetics and algebras associated with the oscillation of the reentering mark (and its numerical analog – the oscillations tied to the square root of negative unity). We discovered that the oscillation of plus and minus leads to the square root of minus one and to Clifford algebra while the oscillation of 0 and 1 (Boolean or numerical) leads to the creation and annihilation algebra for a Fermi particle. It should be mentioned that this work harks back to the beautiful papers of Henri Bortoft (1970, 1971) where he identifies the zero-one oscillation as the condition of an observer who is placed in a condition where he cannot distinguish the whole from the part. It was Bortoft's intuition that this (in the context of Young's double slit experiment) was the nexus and source of the quantum interference. We shall take up this theme in a separate paper. The paper by Basil Hiley (2013) is a description of our ideas in prior incarnations and a remembrance of Bortoft. It can be read as a companion to this paper. Another connection not discussed in this paper is the system of recursive distinctioning (RD) originated by Joel Isaacson (1981). There a graphical structure holds icons that produce new icons in synchronous recursion based on the distinctions between the icons at any given time. Recursive patterns from the simple oscillation to the highly complex are produced. Isaacson, the author of this paper and others, including Bernd Schmeikal (see Schmeikal, 2012) are investigating this RD structure for its internal properties and for its possible applications. In the present paper we found a non-commutative containment calculus for the Fermion, and finally we located the deep Pythagorean relationship that binds nilpotent Clifford algebra to the Dirac equation in the work of Peter Rowlands. The span of our adventure has been wide, and it has been a walk along a very particular and careful path. All the ideas from beginning to end are related to one another. The relationships we have articulated are but a droplet in the further articulation of the possibility of a distinction.

Acknowledgements

This paper is dedicated to the memory of Annetta Pedretti. The references (Kauffman, 1978–2016) document a segment of the author's prior work on laws of form and the concept of the imaginary. It gives the author great pleasure to thank the many people with whom he has discussed these ideas, and most particularly to thank Art Collings, Graham Ellsbury, James Flagg, Bernd Schmeikal, and Joel Isaacson for a long and continuing process of conversations. This research work was supported by the Laboratory of Topology and Dynamics, Novosibirsk State University (contract no. 14.Y26.31.0025 with the Ministry of Education and Science of the Russian Federation).

References

Spencer-Brown, G. (1969). *Laws of form*. London: George Allen and Unwin Ltd.
Bortoft, H. (1970). The ambiguity of 'one' and 'two' in the description of Young's experiment. *Systematics, 8*(3), 221–244.
Bortoft, H. (1971). The whole: counterfeit and authentic. *Systematics, 9*(2), 43–73.
Collings, A. (2017). The Brown 4-indicational calculus. *Cybernetics and Human Knowing, 24*(3-4), 75–101.
Hiley, B. (2013). The arithmetic of wholeness. *Holistic Science Journal, 2*(2), 23–30.
Isaacson, J. (1981). *Autonomic string manipulation system*. United States Patent 4286330.
Kauffman, L. H. (1978a). Network synthesis and Varela's calculus, *International Journal of General Systems, 4*, 179–187.
Kauffman, L. H. (1978b). DeMorgan algebras—Completeness and recursion. *Proceedings of the Eighth International Conference on Multiple Valued Logic(1978)* (pp. 82–86). Boston: IEEE Computer Society Press.
Kauffman, L. H., & Varela F. (1980). Form dynamics. *Journal of Social and Biological Structures, 3*(2), 171–206.
Kauffman, L. H. (1980). *Paper computers*. Unpublished paper.
Kauffman, L. H. (1985a). Sign and space. In *Religious experience and scientific paradigms*: Proceedings of the 1982 IASWR conference (pp. 118–164). Stony Brook, NY: Institute of Advanced Study of World Religions.
Kauffman, L. H. (1985b). *Map reformulation* (A. Pedretti, Ed.). London: Princelet Editions.
Kauffman, L. H. (1987a). Imaginary values in mathematical logic. *Proceedings of the Seventeenth International Conference on Multiple Valued Logic, May 26-28, 1987* (pp. 282–289). Boston: IEEE Computer Society Pres.
Kauffman, L. H. (1987b). Self-reference and recursive forms. *Journal of Social and Biological Structures, 10*(1), 53–72.
Kauffman, L. H. (1987c). Special relativity and a calculus of distinctions. In *Proceedings of the 9th Annual International Meeting of ANPA* (pp. 290–311). September 23-28, 1987, Cambridge, UK.
Kauffman, L. H. (1988). Space and time in computation and discrete physics. *International Journal of General Systems, 27*(1-3), 249–273.
Kauffman, L. H. (1994). Knot Logic. In *Knots and applications* (pp. 1–110) Singapore: World Scientific.
Kauffman, L. H. (1996). Virtual logic. *Systems Research, 13*(3), 293–310.
Kauffman, L. H. (2001). The mathematics of Charles Sanders Peirce. *Cybernetics and Human Knowing, 8*(1-2), 79–110.
Kauffman, L. H. (2002a). Biologic. *Contemporary Mathematics, 304*, 313–340.
Kauffman, L. H. (2002b). Time imaginary value, paradox sign and space. In D, Debois (Ed.), *Computing Anticipatory Systems: CASYS 2001—Fifth International Conference* (Vol. 627, pp. 146–159). Proceedings of a conference held in Liege, Belgium, August 13-18, 2001. Melville, NY: AIP Conference Publishing.

Kauffman, L. H. (2003). Eigenforms—Objects as tokens for eigenbehaviors. *Cybernetics and Human Knowing, 10*(3-4), 73–89.
Kauffman, L. H. (2005a). Eigenform, Kybernetes. *The Intl J. of Systems and Cybernetics, 34*(1/2), 129–150.
Kauffman, L. H. (2005b). Reformulating the map color theorem. *Discrete Math., 302* (1-3), 145–172.
Kauffman, L. H. (2009). Reflexivity and eigenform: The shape of process. *Constructivist Foundations, 4*(3), 121–137. Retrieved September 17, 2019 from https://homepages.math.uic.edu/~kauffman/ReflexPublished.pdf
Kauffman, L.H. (2012a). *Knots and physics*. Singapore: World Scientific.
Kauffman, L. H. (2012b). Categorical pairs and the indicative shift. *Applied Mathematics and Computation, 218*, 7989–8004.
Kauffman, L. H. (2016a). Knot logic and topological quantum computing with Majorana fermions. In J. Chubb, A. Eskandarian, & V. Harizanov (Eds.), *Logic and algebraic structures in quantum computing and information: Lecture notes in logic* (pp. 223–335). Cambridge, UK: Cambridge University Press.
Kauffman, L. H., (2016b). Cybernetics, reflexivity and second order science. *Constructivist Foundations, 11*(3), 489–507.
Kauffman, L. H. (2017). Imaginary values. *Cybernetics and Human Knowing, 24*(3-4), 189–223
Rowlands, P. (2007). *Zero to infinity: The foundations of physics*. Singapore: World Scientific.
Shannon, C. E. (1938). Analysis of relay and switching circuits. *Transactions American Institute of Electrical Engineers, 57*, pp. 713–723.
Schmeikal, B. (2012). *Primordial space: Pointfree space and logic case*. Hauppauge, NY: Nova Science.
MacLane, S. (1986). *Mathematics form and function*. New York: Springer-Verlag.
Varela, F. J. (1975). A calculus for self-reference. *International Journal of General Systems, 2*, 5–24.
Varela F. J. (1979). *Principles of biological autonomy*. New York: North Holland.
Wittgenstein, L. (1922). *Tractatus Logico: Philosophicus*. New York: Harcourt, Brace and Company, Inc.

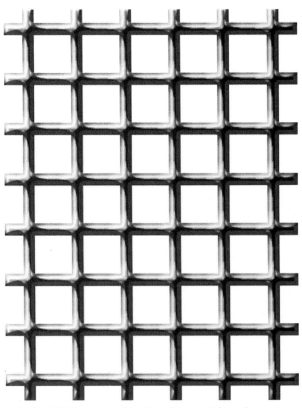

Snelson, P., II. (2009). *Matriarchy: The Ultimate Rubric*. Computer graphic.

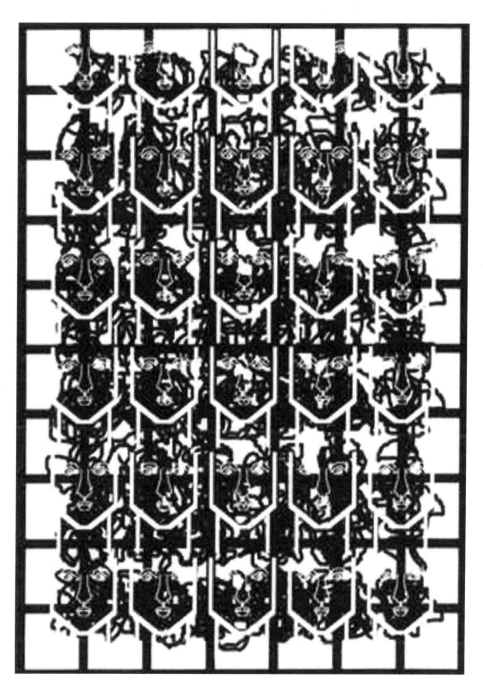

Snelson, P., II. (1993). *Mindsets United*. Computer graphic.

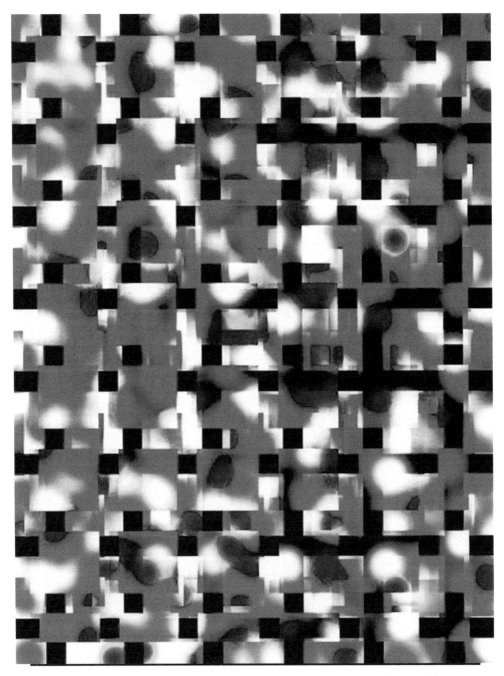

Snelson, P., II. (2001). *Bullets From the Big Sky of Manhattan.* Computer graphic.

Virtual Logic—Who Shaved The Barber?

Louis H. Kauffman[1]

I. The Tale of the Barber in the Frozen Wastes

It is an old tale and a puzzle. In a town somewhere in the wilds of northern Alaska there lives a barber. This barber shaves every man who does not shave himself. An excellent barber he is, and the only barber in his region. People come from miles around to get a shave from him, He has hardly any time for himself.

But it is a puzzlement. Who shaves the barber? It is apparent that he is indeed well-groomed and shaved, and yet the more clever of the townsfolk reasoned as follows: If he does not shave himself then he must shave himself, for he shaves all those who do not shave themselves. This is a contradiction. Therefore he shaves himself. But he cannot shave himself since he shaves only those who do not shave themselves! That too is a contradiction. What is to be done here? The barber is a contradiction. Can he even exist in a logical world?

And in the town lives a curious fellow named Douglas. Douglas has a doctrine that gives him great pleasure, and he loves to tell it to everyone he meets. Douglas would tell you this at the Frozen Wastes Saloon, if you met him there. He would put his arm around you and call the bartender over and say "Give this fellow a good strong whiskey and I'll tell him a tale!" And after a little drinking Douglas would say "You know I will never forget the day I discovered that I have no head!" Now this seems a bit absurd, particularly when you meet Douglas and he has a fine head of hair, fur cap and well-trimmed beard, and the head seems firmly attached to his body as well. So you protest "Seems to me you have as good a head as any man in the Frozen Wastes."

"But, ah" says Douglas, "I did na mean that you don't see me head, but I surely do not see me own head (if I had one which I don't) and what I canna see don't exist for me. Try it yourself. Do you see your head? Where are your eyes? Do you see a mouth beneath nose and all that other fol der rol of a face that I sees over on you? Nope. You've got all that apparatus, comical as it is, but not me!"

And when I met Douglas, I decided to trap him. So I sez "Do you shave or trim your beard?" Douglas answered forthrightly "I shore do!" And I sez "But don't you need to have a head to have a beard to trim and shave?" Well...Douglas surprised me. He said, "I don't shave myself. I shave my mirror image. It is real simple. I have no head and all the rest, but when I look in a mirror I see that fol de rol that you call a head on that image in the mirror, and I gets my razor and my scissors and I gets to work, always looking at that mirror image. I ain't got none of that stuff but I know that

---

1. Email: kauffman@uic.edu

when I keeps my mirror image spruced up, why the ladies they like it better that way. So I shaves and trims my mirror image, but I surely do not shave myself."

What is your profession I say to Douglas. "Why I am the barber in these here parts.

I shaves every man who does not shave himself. Some people around here think I am contradictory and should not exist, but in fact my head does not exist and so the matter of my shaving myself is just not an issue. I would not shave a headless man.

Isn't it funny how people can get all perplexed by a puzzle and it turns out to have such a simple solution?"

So that solved the problem of the barber. The barber, from his own perspective, has no head and so can neither shave nor not shave himself. There is no contradiction in the statement that the barber shaves everyone who does not shave himself. And this particular barber, does shave and trim his mirror image.

II. Return of the Barber

Lets formalize this headless solution to the barber paradox. We can say it this way:

Each person X has a mirror image X* that is accessible to X by looking in a mirror.
X* does not exist if X is blind.
No man M can shave a mirror image X* unless it is his own mirror image.

These statements apply to a person X even if he believes that he has a head. Shaving is an activity that is coordinated by means of a mirror. Of course a blind person does not have a personal mirror image and so we assert that X* does not exist for a blind man, since we are concerned with the personal mirror image.

With this background about mirror images, we give the

New Barber Axioms
The barber shaves the mirror image of X if and only if X does not shave X.
The barber shaves X whenever X does not shave X*.

Since no man can shave a mirror image unless it is his own mirror image, the first axiom really says "The barber shaves his mirror image." And we now know that this is no paradox since no one shaves himself, those who appear to shave themselves actually shave their mirror images.

And the blind man. Well he certainly does not shave his mirror image since there is no mirror image for him. Thus he is shorn by the barber.

III. The Russell Paradox

The famous Russell paradox is a close relative of the barber paradox.

Bertrand Russell proposed the following set:

R is the set of all sets that are not members of themselves.

Conceptually, the Russell set R appears to make good sense. After all some sets do belong to themselves. The set of all ideas is an idea. And some sets certainly do not belong to themselves. The set of all ducks is not a duck.

But what about R itself? If R belongs to itself, then by the very definition of R, R cannot belong to itself. But it R does not belong to itself, then by the very definition of R, R must belong to itself. We are caught in a contradiction either way.

Can we solve the Russell paradox with mirror images? Let us try. For every set X, let X* denote the mirror image of X. We double the world of sets so that whenever you have a set in the usual world, you have a corresponding set in the Looking Glass World. It is just like what happened to Alice in Lewis Carroll's fantasy *Through the Looking Glass* only it is very precise. If I have a farm with 20 chickens and a rooster in this world, there is a mirror me and mirror chickens and roosters in the mirror world. We could then find out about passages back and forth between this world and the mirror world. But this is another tale for another time.

Now we shall have a new Russell set R defined by the statements:

New Russell Axioms
X is in R if and only if X* is not in X.
X* is in R if and only if X is not in X.

Note that we imagine sets that have both standard elements and mirror elements in them and it could even happen that a mirror set was a member of a given set. What would the simplest example of this be? We could have S = {S*} so that the only member of S is its mirror twin S*.

But then S* would have as its member S** and we would have S = {S*} = { {S**}} = {{{S***}}} = ... and it is like looking into two facing mirrors and seeing many reflections at many depths. OK?

Note that we really have constructed a kind of analog of the original Russell set since we place X* in R whenever X is not a member of X.

Now what about R. Does R get into trouble? Here are the statements for R as a possible member of R or of R*:

R is in R if and only if R* is not in R.
R* is in R if and only if R is not in R.

Nicely, these two statements say exactly the same thing. So R is in R if and only if R* is not in R. We have given R a choice, and the choice can be made either way. We can have R a member of itself, but then R* will not be a member of R. Or we can have

R not a member of itself, but then R* will be a member of R. Choice has replaced paradox. This is a happy way to resolve the Russell paradox.

Remark. The thoughtful reader may be wondering if R might now be the set of all sets. Call a set "standard" if X* is not a member of X. Then we see from the axiom

X is in R if and only if X* is not in X

that R contains all standard sets. This means that R is at least as big as the universe of all sets before the mirrors were set up! But a little discussion will help.

Imagine that U is the set of all sets in the usual discussion of sets. Then it is often said that U is self-contradictory because Cantor has shown that any set is less than its set of subsets, but the set of subsets of the set of all sets is just equal to the set of all sets! The set of all sets cannot be greater than itself. But let us look carefully at how Cantor's argument works here. If S is any set such that every element of S is a subset of S then we can map S to P(S), the set of subsets of S, by sending a set X to itself (since it is a subset of S as well as being a member of S). Cantor directs us to consider the set C={X in S | X is not a member of X}. Suppose that C = Y for some subset Y of X. This cannot be, for if Y is a member of C then Y not a member of Y and so not a member of Y = C. And if Y is not a member of C, then Y is not a member of Y and so is a member of C. We conclude that C is not in the set S and so S is seen to be smaller than P(S) with respect to this inclusion. The set C is in P(S) but it is not in S. The argument applies similarly to any inclusion of S in P(S). But if S = U, then C becomes the Russell set C={X | X is not a member of X}. In our discussion we have banished the Russell set from the lands of Setdom. At the point of examining the set of all sets, Cantor's argument loses its traction. Just at the point where our Barber could be in trouble, the absence of his head saves the day! More analysis is required here. We leave the matter to the reader and his faithful mirror.

IV. Condensation and Generalization of the Formalism

Here is a formalism for thinking about the problems we abandoned at the end of the last section.

Let Ax mean that "x is a member of A."

Then the Russell set is defined by RX = ~XX, which says that X is a member of R if and only if X is not a member of X.

The new Russell set of the previous section is defined by

$$RX = \sim XX^*$$
$$RX^* = \sim XX.$$

From this we get that RR = ~RR* and the freedom of choice that ensued in the previous discussion.

We have avoided RR = ~RR and instead have RR = ~RR*. The pattern of this is A = ~B and of course B = ~A.

We have avoided G = ~G by letting G split into A and B with A = ~B and B = ~A. The worst that can happen here is to substitute ~A for B in A = ~B and obtain A = ~~A. We do not regard such double negations as problematical!

There is even a way to obtain a fixed point for negation from the pair (A,B). Define ~(X,Y) = (~Y,~X). Then ~(A,B) = (~B,~A) = (A,B). So we could take G = (A,B) and assert that ~G = G. But the logic of such pairs is a bit different from the Boolean and we shall forgo an exploration of that direction in this essay (see Kauffman, 1978)).

V. Epilogue

The singularity in logic initiated by the Russell set has been told again in this essay via a parable of the Barber who shaves every man that does not shave himself. But this Barber avoids the paradox of his own shaving by not having a head! (See Harding [2013] for the source of this idea of headlessness.) In not having a head he has no responsibility to shave himself or not to shave himself. In fact, he shaves his mirror image and so remains acceptable to the social world. We have suggested that a similar solution to the Russell paradox is available where the Russell set contains mirror copies of all sets that do not have themselves as members, and indeed the new Russell set could take on its own mirror image. It has that choice, just as the Barber has the choice to shave his mirror image. When the new Russell set takes on its mirror image as a member, then it is not a member of itself and so becomes a socially acceptable set in the community of those to whom self-membership is anathema. Perhaps the reader as cyberneticist would like to draw further conclusions from this tale of external, internal and mirror worlds.

References

Harding, D. E. (2013). *On having no head*. London: The Shoreland Trust.
Kauffman, L. H. (1978). DeMorgan Algebras—Completeness and Recursion. *Proceedings of the Eighth International Conference on Multiple Valued Logic* (pp. 82–86). Boston: IEEE Computer Society Press

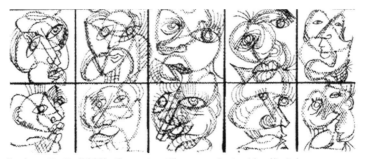

Snelson, P., II. (2009). *Capturing Character Series* (detail). Ink on canvas.

Snelson, P., II. (2019). *Cybernetic Art Matrix Revitalized* (vertical). Computer graphic.

ASC
American Society for Cybernetics
a society for the art and
science of human understanding

The Destabilizing Cybernetics of Implausibility:
The Anti-Anthropocentric Crisis

Zane Gillespie[1]

1. Introduction: Implausibility

On the day after a panel/plenary discussion signaling the centenary of Herbert Brün's birth, I presented a lecture/recital as part of the American Society for Cybernetics (ASC) 2018 Conference. It featured my performance of *The Human and Non-human for Piano and Digital Delay* (2018) (*HNh*), a set of 11 short pieces. One distinctive aspect of this piano music is the neo-Romantic harmonic fingerprint that has heretofore defined my musical personality. And yet it is full of tell-tale signs of my indulgence in a sort of clandestine protest against that backward-looking style. To audition my performance of *HNh* at the ASC 2018 Conference, please click on the following link: https://www.youtube.com/watch?v=ryr36AXWwF0.

I composed *HNh* after reading a paper on posthuman creationism by David McWilliam, Lecturer in Film Studies and English Literature at Keele University in Staffordshire, England. Posthuman creationism is the idea that all living organisms (including human beings) are "mere biological matter awaiting further transformation" (McWilliam, 2015, p. 541) originating from specific acts of posthuman creation, rather than by natural processes such as evolution. This posthumanistic theory—along with any other theory, philosophy, ideology, movement or view that seeks to extend the human species beyond itself—is likely to seem highly implausible to most people if only because it represents an affront to human dignity, an inadequately defined concept which holds all that is idiosyncratically human as "something sacred, inviolable or essential" (Palk, 2015, p. 39).

---

1. Executive Board Member for Music Theory for the Southern Chapter of The College Music Society.
   Email: rzgllspe@memphisalumni.org

However complex a concept like human dignity, anything offending against it (and so perceived to be implausible or, shall we say, anti-anthropocentric) must be at least as complex. Indeed, implausibility is a complex subject. Brün's (2004a) exposition of the relevance of implausibility offers a logical explanation for the "inconceivable" work of composers who, in the interest of creating something new, produce unfamiliar musical events. As he explains, apologists had long tried to convince people that implausible works of art were actually not that implausible. In the case of the work of Austro-Bohemian composer Gustav Mahler (1860–1911), however, late-19th-century audiences found it implausible because of its extreme plausibility. They attacked its plausibility because "it seemed to tell them with superfluous insistence what they believed they already knew" (Brün, 2004a, p. 96). With the work of Austrian (and later American) composer Arnold Schönberg (1874–1951), it became clear that logical structure does not necessarily result in plausibility. This led to a bias against the so-called intellectual composer, while simultaneously supporting his devotees' contention that understanding a composition's logical structure will make it plausible. According to Brün, however, the extreme plausibility of verbal language makes this impossible. Indeed, composers since the mid-20th century have wondered whether plausibility conduces to understanding at all. Consequently, no one concerns themselves anymore with whether or not their work is plausible. The problem of how to impart anything educative about implausible art, therefore, remains.

This paper proposes a cybernetic interpretation of implausibility. By proposing definitions and models, this paper will lay the groundwork for an anti-anthropocentric compositional theory (ACompT), based on the idea that implausibility is directed by a *feedback loop*, which drives the production of anti-anthropocentric art, a typology of which is given in section 3. It will also examine the possibility of a heuristic application of this ACompT to the theory and analysis of art. The focus, however, will be on *HNh* and the feedback loop that is purported to function internally to it.

Definition 1. Implausibility: For the purposes of this paper, an implausibility will be defined as a plurality of things–simultaneously interconnected in network relations and differentiable—which, individually, produce no disruption of human thought (i.e., the goal-oriented flow of ideas and associations leading to human-oriented conclusions about reality). When considered together, however, this plurality seems hardly likely to be true, and produces an emergent challenge to human hypocrisy and complacency, which is not generated by the components themselves, but by the unlikelihood of the composite.

Implausibility is observable (e.g., the apologia for the possibility of silent music that is John Cage's *4'33"*), or simply imaginable (e.g., Nam June Paik's "Danger Music for Dick Higgins," which directs the performer to "creep into the vagina of a living whale"). In any case, if considered objectively, implausibility appears distinct with respect to that of which it is composed and thus, seems to show an anti-anthropocentricity that is autonomous of its individual elements yet due to the composition (i.e., putting together) of those elements.

The analysis and understanding of the phenomenon of implausibility in art are made even more difficult because people tend to repudiate implausibilities, and act only to avoid cognitive dissonance. Rather than change their standards according to dissonance-reducing cognitions of dissonance-increasing information, they seem to follow some necessitating, universal, and self-preserving instinct (Festinger, 1957). This instinct derives from some primitive hostility triggered by implausibilities (e.g., the composition whose instrumentation calls for vacuum cleaners and low-energy light bulbs, the opera without a plot, the piece with three different ensembles separated from each other and playing in wildly different rhythms and keys, the use of unorthodox instrumental techniques, any music listening requiring effort, etc.). Otherwise, how do we explain the common perception of modern art as a rejection of traditional values? Why has there always been a negative correlation between new art and a mass audience (Holbrook, 2012)? What romanticism encumbers the minds of those who regard modern art as intentionally destructive? Are untrained audiences the most appropriate for new art? How do we explain conventional art's characterization as bourgeois? What can explain the widespread prejudice against modern art? Why is modern art so often perceived to be psychopathic and as having no inherent artistic value (Scott, 2009)? From whence comes the bourgeois notion of music strictly as an escape in entertainment, reducing music to a kind of drug? Why have so many artists sought to subvert the ideological foundations of mainstream culture? How do symphony orchestras justify the exclusive programming of works which guarantee maximum bourgeois decorum? Why do most people expect to engage music passively?

It is important to note explicitly that, from Definition 1, it follows that implausibility is identified with composition (i.e., putting things put together to create a work of art). Human society is the system in which certain composers (i.e., artists, including composers of music) live (although they try to distinguish themselves from this society), think (although they may never understand why they are misunderstood), and work in rebellion against patterns of stable social relations of co-existence imposed on them by systems of authority. This rebellion manifests as uncooperative behavior that is disdained by most members as inessential to the survival of human society even while it results in the creation of works of art. In human society, while its members share their cultural norms, their general interest (the common good), and their expectations for behavior, these are largely spurned by those who do not believe in "consistency, coherence, communication, perfect models, and other such comfort-providing, distinction-removing ... paradigms ..." (Brün, 2004b, p. 225).

Instances of implausibility in art are delimited by a structural quality causing them to be perceived by the mainstream as unstructured artifacts of a misanthropic, sociopathic mind with no interest in helping to achieve the common goals or the common good of human society. Take this scathing invective against Mahler's music, for example:

So, even to those whom it does not offend directly, it cannot possibly communicate anything. One does not have to be repelled by Mahler's artistic personality in order to realize the complete emptiness and vacuity of an art in which the spasm of an impotent mock-Titanism reduces itself to a frank gratification of common seamstress-like sentimentality (Louis, 2000, p. 4).

2. Anti-Anthropocentrism: A Feedback Loop

Definition 2. Anti-anthropocentrism: For the purposes of this paper, anti-anthropocentrism will be defined as a composer's conscious or unconscious putting together of an implausibility's (see Definition 1) component parts as a negation of humanist values, which they must still directly engage as a consequence of their own humanity. On the one hand, anti-anthropocentrism is, or originates from, the arbitrary phenomenalization of the human being (represented by the implausibility itself), which derives from the anti-humanism of the composer; on the other, this phenomenology affects the composer's human consciousness and becomes in their work an expression of a continuously changing, multi-perspectival, open-ended process of surpassing our evolutionary heritage.

According to Definition 2, anti-anthropocentrism is bound, ironically enough, to the humanity of the composer. Yet defining the human being phenomenologically does not necessarily lead to the notion of transcending our evolved traits to become something other than human; it becomes a factor in this notion only when the human mind is regarded as disembodied data that can be shifted from one body to another not-necessarily-human one. Likewise, it is given an equally neo-idealist interpretation by the composer for whom it is a substrate for their creative decisions, which shift from one form to another throughout a work, and from work to work. For this reason, anti-anthropocentrism can also be called anthropogenic[2] anti-humanism to distinguish it from non-anthropogenic anti-humanism. These are two extremes of a spectrum in which one antipode is lost (by degrees) as a discrete entity to become the other. This is due to a feedback loop which shares an identity of essence with that which produces the perceptual continuum between the human and the non-human. As an analogical model, one might consider the feedback loop involved in the self-organizational flow, as it were, between disordered and ordered living systems (Beer, 1994). Because, when considered abstractly, composers are similar and do similar things, it follows that we can assume that any anti-humanistic (viz., anti-anthropocentric) urges they have will produce similar creative products which, paradoxically, are simultaneously anthropogenic and therefore based in the humanistic foundation of artistic discourse (Pop, 2014; Burke, 2008). A typical example of humanism by way of anti-humanism (and vice versa) is any exceptionally individualistic work of art (Fox, 2013).

I propose the following:

---

2. This adjective is used here to mean originating in human activity and not to pertain to the study of the origins of humanity.

Definition 3. Feedback loop: In an implausibility, the mutual dependence between a composer's humanism and the implausibility's ostensible anti-anthropocentricity (and its intended effects) can be defined as a feedback loop.

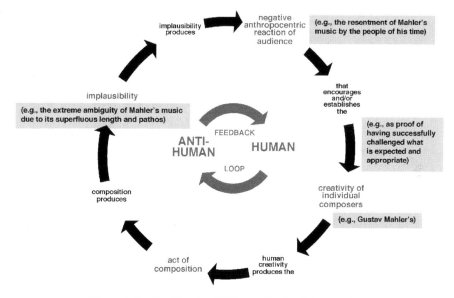

Figure 1. Logic of implausibility as a feedback loop system

The feedback arises and is maintained by a set of necessitating factors, which impel the composer to adapt their humanity to an expression of anti-humanism; the feedback is maintained by the action of an anti-anthropocentric creative urge, which leads to the implausibility to negate humanist values to produce and maintain an anti-humanist opposition to those values. A necessitating factor is defined as anything—a constraint, control, rule, impulse, condition, compulsion, need, and so forth—obliging the composer to create the implausibility to adapt their humanity to an expression of anti-humanism. In this sense, it is both the ἐνθουσιᾰσμός (enthousiasmós) of the Greeks and Romantics as well as something akin to the materialist idea of inspiration as a result of the artist's unique reception of the signals, as it were, transmitted from an external crisis. The stronger the necessitating factors, the more this kind of artistic creativity is activated. In anti-anthropocentric art, the necessitating factors often originate from conscious motivations: the desire to criticize traditional bourgeois values, to dislodge idealist concepts, to transcend the humanist paradigm, to shock, and so on. At other times, these factors can also be natural and act unconsciously, as they may derive, at least in part, from a genetic predisposition toward non-acceptance of cultural norms and anti-authoritative attitudes. The existence of one or more necessitating factors is indispensable though not sufficient; it is also necessary for the process to decentralize and produce the opposite of whatever is culturally sanctioned. Through the feedback loop, this becomes a necessitating factor itself and so influences the creativity of individual composers. This anti-anthropocentripetal logic can be represented as shown in Figure 1.

Recognizing the existence of a feedback loop and understanding the nature of necessitating factors is indispensable for interpreting implausible phenomena.

Here is an example: If a composer, non-randomly, decides to eschew all pattern and repetition, listeners are unlikely to enjoy their inability to find pattern and repetition, and, because of the evident presence of an anti-anthropocentripetal system (see Definition 2), must be convinced of its plausibility (Brün, 2004a); however, if a composer writes music familiar to and popular with the majority of people in their culture, the new and unfamiliar may still be expressed, but not necessarily with the need to invoke an anti-anthropocentripetal system; listeners will simply digest the new by measuring it against what is familiar.

3. Typology of Anti-Anthropocentrism in Art

The author has ordered and classified anti-anthropocentrism in art into five types.

3.1 Nihilistic Anti-Anthropocentrism

Nihilistic anti-anthropocentrism, whose anti-humanism manifests by shirking the problem of meaning altogether; this shirking represents the malicious assertion made by certain composers through their art that life is meaningless; if news media, critics, and composers have a taste for art that contributes to the denigration of humanness, they are necessarily engaged in an assault on human reason, although they may only be looking for an advantage in the world of what is fashionable at the moment (Holbrook, 1980). This applies to a diverse range of phenomena, among which is the suppression of critical discussion, the government subsidization of "degenerate"[3] art, and the creation of entirely "negative" art, and in which ontological truth and causality no longer matter. It can also be applied to phenomena such as art based on the inversion of accepted moral standards, and in which the real presence of the real human body is jeopardized by various other virtual presences (Virilio, 2003).

3.2 Biotechnological Anti-Anthropocentricism

Biotechnological anti-anthropocentricism is the combined use of current methodologies of art, biology, and genetics by artists working with genetic technologies involving bacteria, plants, and animals; their experience-based inquiries are not driven by hypotheses but by their stated aim of creating new life forms (the implausibilities, in this case). This art practice is also meant as a critique of the implications and outcomes of genetic technologies; achieving these ambitions, however, seems to militate against accepted ethical standards. Biotechnological anti-anthropocentricism gives rise to a diverse range of phenomena: from the integration of human genes in an artwork to the use of semi-living skin-, muscle-, and bone-tissue cultures as the artist's materials, from transgenic art, involving the transfer of genetic material to create hybrid organisms, to releasing them into the environment. These

---

3. The author indicates irony when introducing specific instances of terminology to evoke questions of what is being said, certainty, and to indicate an epistemological and/or affectational shift.

phenomena are the product of an aesthetic that is oppositional to anthropocentric biases concerning the manipulation of living systems (Gigliotti, 2008).

3.3 Ultraterrestrial Anti-Anthropocentrism
Ultraterrestrial anti-anthropocentrism is the creation of art, as if the artwork itself, as a liminal space between the imagination and the senses, could give a narrative, visual, or aural representation of culturally divergent, ultraterrestrial beings (to borrow a term from journalist and parapsychologist John Keel [2002]). However, the mythological, folkloric, or paranormal figures, at once anthropomorphic and inhuman, are the inspiration for the composer's engagement with cross-cultural, fairy-type stories of archetypal beings. This kind of anti-anthropocentrism has resulted in quite an array of artworks: from composer Claude Debussy's "Sirénes" from his *Nocturnes for Orchestra* (1899) to Susan Hiller's sound-sculpture installation *Witness* (2000); from the literary corpus of H. P. Lovecraft (1917–1937) to the draftsmanship of H. R. Giger (1960's–2014); from "Little Gidding" from poet T. S. Eliot's *Four Quartet's* (1941) to Einojuhani Rautavaara's orchestral work *Angels and Visitations* (1978).

3.4 Posthuman Anti-Anthropocentrism
Posthuman anti-anthropocentrism in art, which assumes *homo sapiens'* biological limitations, pertains to the composition of works that are humanly impossible to faithfully perform, execute, or appreciate. Inevitably inaccurate attempts to realize such works are due to the difference between human capabilities and the transcendent powers of the posthuman being. Posthuman anti-anthropocentrism can be used to interpret a large number of phenomena: from works that are not humanly sensorial (e.g., conceptual artist Joseph Kosuth's *The Second Investigation* [1968] and post-conceptual media artist Maurizio Bolognini's *Sealed Computers* [1992–97]); to works which either parallel or prefigure that which is beyond human comprehensibility (e.g., conceptual artist Ian Burn's *No Object Implies the Existence of Any Other* [1967] and the new-media art in the exhibition *Les Immatériaux* [Centre Pompidou, 1985]); to works too fast for any human to play (e.g., György Ligeti's *Étude No. 14A: Coloana fara sfârşit* [*Column without End,* 1988–1994]); to works significantly longer than any human performer's lifespan (e.g., Bull of Heaven's *310: ΩΣPx0(2^18×5^18)p*k*k*k* [2014]).

3.5 High-Technological Anti-Anthropocentrism
High-technological anti-anthropocentrism has to do with the movement of artificial intelligence into the domain of human creativity, understood as the machine synthesis of inputs to generate original works of art, by means of automated, algorithm-driven processes. In particular, this offers a simple explanation for the direct proportion between the explosion of artificial creativity and the continuous, exponential, and irreversible increase in the technologies that have already revolutionized our way of life.

3.6 Psychological Anti-Anthropocentrism

Psychological anti-anthropocentrism refers to the equivalence between all that is anti-anthropocentric and everything the conscious human ego refuses to identify in itself. Because one tends to reject or remain ignorant of whatever one finds objectionable, anything that challenges whatever one finds acceptable is viewed negatively. It is the unknown, humanity's greatest fear. Thus, we insulate ourselves by constantly raising a shield of illusion between the ego and the real world. In one sense, the anti-anthropocentric is roughly equivalent to the Jungian *shadow* archetype (Jung, 1969); the result of Jung's elucidation of this unconscious aspect of the personality can be interpreted as an elaboration of the human-centered belief that evil things exist in nature. In spite of this, the anti-anthropocentric, like the shadow, may be a latent source of potent creativity. Consequently, the most interesting anti-anthropocentrism may be one's own. Take, for example, the present author's own work *The Human and Non-human* (2018), the composition of which resulted from a personal recognition of the inability of the human mind to correlate all the information it contains and the concurrent intuiting of the insignificance and powerlessness of humanism against the world's implacable, anti-human bleakness. Implausibility in art, by virtue of cybernetic behavior and even as it evokes this cosmic inhospitableness, is the inverse instantiation of, the negative space around and between, or, one might say, the shadow of humanism (see sections 1 and 2).

4. The Human and Non-human for Piano and Digital Delay

Definition 4. The human and non-human for piano and digital delay (HNh): *HNh* is an anti-anthropocentric composition-cum-pedagogical device by the author, which juxtaposes anachronistic, "humanistic" conventionalities (for which the piano is a metaphor) with an "anti-humanism," associated with an ever-present, "techno-scientific," digital-delay effect. Each of the composition's 11 movements denotes a stage in a progression meant to gradually eliminate the gap between the human and the non-human, or, conversely, to increase the sense of a single identity shared by both. Although an anthropogenic experiment, the purpose of the constant updating of variables (e.g., delay time, sine-wave modulation, etc.) is to incrementally unveil an "unconventional," "non-human" music, according to typical feedback-loop action.

Each movement of *HNh* can be described by functioning rules (that also define the work as a whole). These rules are an account of the liminal space in which there were both periods of clarity and times of struggle in the creation of *HNh*. They represent the moment between the impulse and the action of composition, and are as follows:

1. Each movement of *HNh*, is composed of a *base*, of *A* (nomic; conventional; accessible) "human" music, M_a, to which is applied a digital delay effect, consisting of delay time, feedback, and sine-wave modulation.

2. These individual variables and their values represent the state of M_a for each movement, x_n, $n = 1, 2, ..., 11$.

3. Thus:

$$HNh = (x_1, x_2, \ldots, x_{11}) \qquad (1)$$

defines the state of each M_a for each movement.

4. Let us imagine for each M_a, a function, p, transforms the state of each movement, x_n, into an effect, e_n, which can be conceived as the movement's output:

$$p(x_n) = e_n \qquad (2)$$

Here, the use of the word *effect* refers to the ability of art to induce changes of mood and deeper levels of thought. In the case of *HNh*, it was intended that the perception of each movement over the course of its performance time would be something the percipient might reasonably interpret as symbolic of a gradual destabilization of our anthropocentric narrative of the universe.

5. Let us imagine that, due to the authorial intent of unifying the human and the non-human, there is a function, u, which allows *HNh* to unify the dual nature of each movement, x_n, to produce a synthetic state, s_n:

$$u(x_1, x_2, \ldots, x_{11}) = S \qquad (3)$$

Once again, the intended perception of the synthetic state of each successive movement over the course of *HNh*'s performance time is one which listeners might reasonably interpret as symbolically militating against the propositions of philosophical humanism.

6. Finally, let us imagine a function, g, which transforms each successive movement's synthetic state into a processually synonymical iteration of the last, and represents the goal-direction for *HNh*:

$$g(x_n, s_n) = x_{n+1} = \text{goal-direction for } HNh \qquad (4)$$

Equation (4) is also necessary to specify how the "accessible," "human"-music bases of each movement at x_n are linked to the superimposed, "anti-humanistic" entropy of *HNh* at x_{n+1}. Furthermore, it shows that the execution of the intention to increase each movement's "listener-unfriendliness" at $xn+1$ is a function of the global goal-direction for *HNh* and so intersects with the "listener-friendly" bases at x_n as well.

7. The feedback loop is described by the following system:

$$\begin{cases} u(x_1, x_2, \ldots, x_{11}) = S \\ g(x_n, s_n) = x_{n+1} \end{cases}$$

(5)

There is, at least philosophically, a one-to-one correspondence between how a composer executed his intentions (e.g., how the present author wrote each movement according to a process of functional iteration—a feedback loop where each movement is the output which serves as the next movement's input) and what the listener hears during a performance. Suppose the intended listener

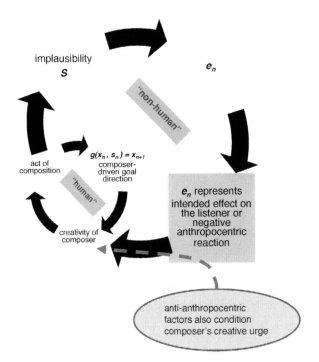

Figure 2. Schematic general model of *HNh*

reaction to a movement was meant to depend entirely on the intended listener reaction to the last. A composer could imagine a table listing all the specific possibilities—that listening to one movement is not intended to be terribly hard means that listening to the next will be intended to be a little harder, and so forth. Or this same composer could capture (as the present author has done) the relationship of the intended degree of difficulty of one movement to that of the next as a rule, as a function. Especially when combined with Equation (3) in the System (5), one sees that the intended listener reaction of the next movement (x) is a function (g) of the intended listener reaction of the last: $g(x_n, s_n) = x_{n+1}$. System (5) is illustrated in Figure 2, instantly giving a sense of the very close correspondence between the composition of each movement and how each movement is intended to be received by the listener.

8. It was intended by the composer that the aspect perceptional measure of the composition as a human enjoyment should gradually lead to a shift of the aspect perceptional measure of the composition as normatively unenjoyable, following the logic shown in Figure 3, which represents the operative logic of *HNh* as a self-referential model.

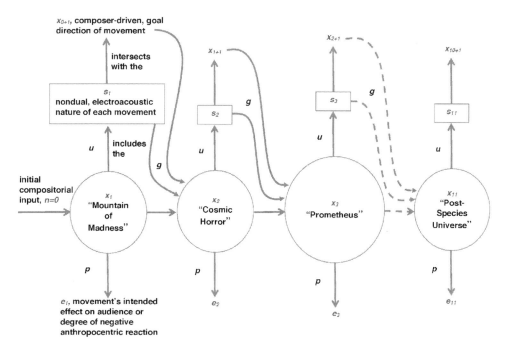

Figure 3. Operative logic of *HNh* as a self-referential system

9. The functions *u* and *g* are the fundamental rules of the System (5) and describe the feedback-loop behavior, as clearly shown in Figure 3:

   • the rule *u*—defined as synthesizing—shows how the individual variables of each movement combine to permit *HNh* to cultivate the effect it should have on the listener, which may be conceived, or interpreted, as increasingly non-vernacular; and
   • the rule *g*—defined as recursive—determines the individual variables of each movement as a function of the internal, increasingly non-vernacular effect, considered to be the authorial-intentional input for compositional decisions which self-referentially change the values assumed by each movement's variables, which should, in turn, modify any perceived human/non-human gaps, driving *HNh*'s evolution over the course of a performance.

10. Figure 3 clearly shows that the first movement at x_1 is due to an initial input modifying the nomic base and giving rise to a small degree of anomism; composition begins when the first input is produced by the composer (as in *HNh*) or by established algorithms, as happens, for example, in software programs for machine creativity using traditional rule-based (symbolic) AI, evolutionary algorithms, and so-called deep learning.

11. Finally, a listening audience should be able to observe that *HNh* operates when there is a minimal change (activation) in each movement's individual variables, and increases its anomic aspect toward micropolyphony consisting of the original piano input plus a maximum number of densely canonical audio output signals (textural saturation) caused by the digital delay.

5. The Anti-Anthropocentric Attractor

The general climate in which anti-anthropocentrism flourishes is the one in which composers live, think, and work, and in which they are hard pressed to verbalize the fundamental elements necessary for understanding the operative logic of compositions producing observable implausible phenomena and, above all, how the feedback works. Underwhelming are the attempts to simulate a composition's dynamics by stating—or constructing ad hoc—a set of standards addressing:

- what music means according to human standards;
- the "romantic" struggle of composers to gain recognition for themselves and their works in human society;
- the question of whether they and their works are to be brought to ignominy, or to be carried by popularity to the empyrean; and
- whether to express their primitive hostility to the dominance hierarchies (shared by both humans and non-human primates alike) by tactics of opposition, or by taking a more socially acceptable route.

The problem of induction (Harris, 2017, chpt. 1), however, raises doubts about the validity of the general belief that places all art in a humanistic frame. We have seen that this is a rather crude expectation of uniformity that is liable to be misleading (Russell, 1961). Furthermore, it has been the author's experience that today's climate seems to be characterized by an unspoken presupposition that any attempt to form new patterns is evidence of a stubborn and unjustified contempt toward certain groups of people (e.g., those a supposedly anti-democratic composer would not expect to possess exceptional talent or natural ability), as though in the context of a larger system of subordination (e.g., Eurocentrism at the expense of other cultures). Militating against this paradigm is what the author terms the anti-anthropocentric attractor.

Definition 5. Anti-anthropocentric attractor: The anti-anthropocentric attractor is an attractor (not in the strict mathematical sense) symbolizing the tendency toward anti-anthropocentrism in art. It represents—at any given instant in an artwork—the shift toward and/or the appearance of events that seem deliberately contrary to what human beings normally prefer, corresponding to a state (or states) of unbearable irony. The anti-humanism, paradoxically spurred by human cultural activity, destabilizes both the self-righteous moralization of pancultural conservancy as well as the initiation of human intellectual achievement. Both have become a commercially supported pretense to culture by which Europe and the Western world, over the last 100 years, has inhibited the advent of an artistic singularity.[4] This, in turn, has provoked the reactionary conviction that the sharp discontinuation of a long-established concept will bring about new continuity—as well as the rise of things like constructive (as well as deconstructive) post-modernism.

When considering the modus operandi of the anti-anthropocentric attractor, it is useful to point out that it may epitomize:

- *cognitive dissonance*, if the anthropocentric response of most is to attack an artwork for disturbing a cultural situation, or to rationalize it, or to ignore it, or even to deny it has anything to do with culture;
- *aphasia*, if difficulties of understanding occasion the inference of difficulties of communication, in which case, people are quick to repudiate the cause of those difficulties, namely, the offending artwork; and
- *counterintuitive causation*, if, in an artwork, the concepts of development or evolution, together with adherence to the counterfactual conditional principle of cause and effect, is not just hidden, but entirely absent.

These properties emerge when, to compose the work, the necessitating factors (which impel the artist to create) are not translated into something easily understood (or understood at all) by human beings. In effect, applying platitudinous ideas like "music is the universal language," propagated by a supercilious humanism, to such a composition's characteristics observable, or merely imaginable, should make a consumerist societal interpretation improbable if not impossible. Implausibility therefore exposes the crisis caused by a gross association fallacy. Syllogistically, this fallacy can be expressed thusly: Humans communicate with each other. Humans make art. Therefore, art is a form of human communication.

Implausibility, $I = \{i_n\}$, is different for each anti-anthropocentric composition, and may be:

---

4. Much like the technological singularity, the artistic singularity is the author's hypothesis of what art could be in a world in which commercialism does not inhibit creativity. The implication is that artistic creativity would explode in uncontrollable and irreversible growth, resulting in reciprocal and as yet unimaginable changes to art and human civilization.

- *corrective* if, for instance, $i_n \equiv$ the implausibility of making the equally gross category mistake of making human comprehensibility necessary to something that cannot properly be assigned to things capable of human communication;
- *anti-subjective* if $i_n \equiv$ the implausibility of regarding each percipient's particular experience of the anti-anthropocentric artwork as the ultimate litmus test, determining everything about the work by what it suggests to them;
- *defamiliarizing* if $i_n \equiv$ the implausibility of a truly anti-anthropocentric work of art, either in part or in the whole, suggesting to the percipient, especially the well-educated percipient, "the 'familiar' experiences of which it might seem to be a variation, a rudiment, a caricature or any kind of derived consequence" (Brün, 2004a, p. 98); and
- *post-human* if $i_n \equiv$ the implausibility of the composer attempting to convince their opponents of the anthropocentricity of their work.

Implausibility can act in two ways:

1. As *suppressive implausibility*, in the sense that an implausibility disallows the self-indulgent use of the litmus test of personal subjectivity; otherwise, it is not implausible and so lends itself to human understandability.

2. As *nonlinguistic implausibility*, in the sense that an implausibility exposes the frailty of human language; otherwise, it does not render the human explanation uninformative and so is not implausible.

Putative familiarity—accounting for our unquestioning assumption that art is a medium of human communication—is paradoxically the very thing which prohibits any further inquiry into the unknown; this familiarity will be referred to as ordinary aesthetic framework recognition: Δ.

With the anti-anthropocentric attractor, a special feedback loop concerns the composition's analyzability A and its implausibility I: the inscrutability of a work of art depends on its implausibility, but this implausibility is, in turn, a function of the composition's inscrutability.

The definition of the anti-anthropocentric attractor with feedback-loop action is summarized in the following formal model (6), which includes the elements in Figure 4:

$$\left\{ \begin{array}{c} \Delta \equiv \text{ordinary aesthetic framework recognition} \\ A \equiv \text{composition's analyzability} \\ u(X) = S \text{ synthetic state of composition} \\ g(x_n, S) = x_{n+1} \text{ "processual synonymy" of composition's elements (goal direction)} \\ I \equiv \text{implausibility} \\ p(X) = e \text{ intended effect of composition on percipient} \\ \text{Set: } (\Delta, A, u, g, I, p) \text{ operative program} \end{array} \right.$$

(6)

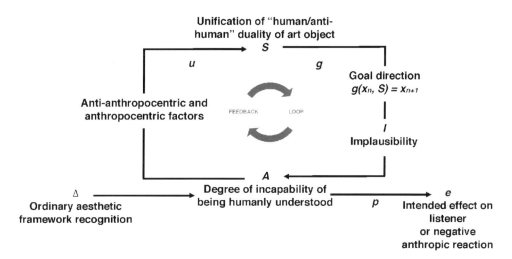

Figure 4. Structural model of anti-anthropocentric attractor [see Figure 3 and System (6)]

The set of rules specifying the composition's identifiability as an aesthetic framework, Δ; its analyzability, A; its implausibility, I; and the functions u, g, and p represent the operative program, which produces the dynamics of the anti-anthropocentric attractor.

The anti-anthropocentric compositions that are perhaps the most interesting are the ones that only seem aphasic. In *HNh*, both the human and non-human elements produce perceptual effects whose implausibility is arguably neither suppressive nor nonlinguistic.

Examples of truly aphasic compositions are those beggaring description, whose repertoire of various elements cannot be recognized as versions of a single base case, although such recognition is typically crucial for the visual perception of objects and events and the audible perception of music (Davies, 2010). In aphasic compositions, while people may recognize easily that a musical framework, for example, is being heard, they are yet helplessly incognizant of its actual existential significance.

Chaos arises in anti-anthropocentric compositions when maximum rationalization (e.g., the correlation of increasing disorder with increasing information epitomized by total serialism) is introduced (Boulez, 1991). These compositions generally instantiate nonlinguistic implausibilities, with the anti-anthropocentric attractor running in the inner recesses of their own apparent randomness. Under certain conditions concerning implausibility, chaotic dynamics can also occur where, regardless of the analyzability, A, of a work's teeming minutiae, audible differentiations on smaller scales are suppressed (Griffiths, 1981). Thus, only a philosophical comparison can be drawn between the work itself and the principle of self-imbeddedness (i.e., similarity across scale), as it is devoid of the exact sort of scaling we find in turbulence (Gleick, 2008).

The laws of probability, as applied to the randomness found in certain complex compositions, may serve not only as a means of analytical approximation but may also

184   Zane Gillespie

explain the rise of emergent properties displayed by those compositions (Prigogine & Stenger, 1984).

6. Digital delay in *HNh* as externalization of the anti-anthropocentric attractor

A composition producing increasingly complex polyphony per digital delay can be seen as an outward manifestation of the attractor whose basic variables and parameters are summarized in the following operative model (7) and represented in Figure 5 which clearly shows the cybernetic processes of the attractor:

$$\begin{cases} \text{for } d_n, n = 1,2 \ldots, N \text{ stored-audio output signals} \\ \text{for } x_n, n = 1,2 \ldots, 11 \text{ initial input signals} \\ (d_n(x_n)) = \text{quasi-canonical, polyphonic texture of input/output} \\ ((1/N)\sum_{1 \leq n \leq N} d_n(x_n)) = D(x_n) \text{ collective layers (cacophony)} \end{cases}$$

(7)

In model (7):

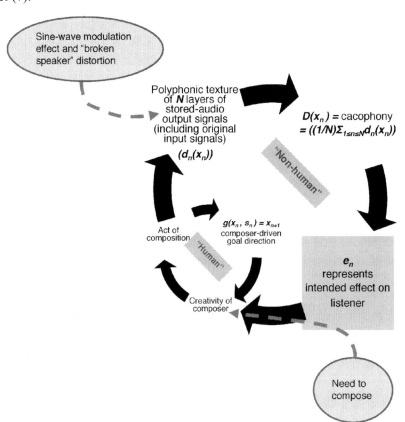

Figure 5. Descriptive model of cybernetic processes in a composition employing digital delay, among other things. (see Figure 2)

1. The polyphonic texture consists of N layers of both stored-audio output signals and original input signals, D_n, with $1 \leq n \leq N$; their behavior is observed over the course of each movement: x_n, $n = 1, 2, ..., 11$.

2. $D(x_n) = ((1/N)\sum_{1 \leq n \leq N} d_n(x_n))$ represents the cacophony arising from increasingly mounting layers due to increasingly longer delay times. We assume this is determined by the arithmetic average of the number of layers.

The preceding rules have been translated into an attractor of N layers, each of which is characterized by individual parameters, which are specific to each movement. These variables are represented in Table 1.

Table 1: The individual parameters for each movement of *HNh*

		I. "Mountain of Madness"	II. "Cosmic Horror"	III. "Prometheus"	IV. "Fragile Mortality"	V. "Foucault"	VI. "The Other"
Title:							
Conventional-music base	M_a Relative intensity: 1-3	2	2	1	1	1	1
"Broken speaker" distortion	Approx. % of wet/dry mix	12.5	12.5	12.5	12.5	12.5	12.5
Delay time	msec	256	256	512	512	256	256
Feedback	%	0	25	50	75	50	75
Sine-wave modulation	Hz	0	0	0	0	0	0

Table 1 (cont'd): The individual parameters for each movement of *HNh*

		VII. "Third Law"	VIII. "Entities"	IX. "Black Seas of Infinity"	X. "Post-Human Creationism"	XI. "Post-Species Universe"	Mean
Title:							
Conventional-music base	M_a Relative intensity: 1-3	1	2	3	2	3	1.73
"Broken speaker" distortion	Approx. % of wet/dry mix	12.5	12.5	12.5	12.5	12.5	12.5
Delay time	msec	256	256	1024	1024	1024	512
Feedback	%	75	75	75	75	75	59.09
Sine-wave modulation	Hz	.1	.1	.1	.1	0	.04

The attractor's externalization is presented in Figure 6, which shows the variables (colored stacked columns) and the growing cacophony (topmost turquoise blue stacked column) over a period of 11 movements (each movement lasts an approximate average of 3' 04"). During the performance of *HNh* at the ASC 2018 Conference, the cacophonic polyphony was reinforced with a soupçon of "broken-speaker" distortion.

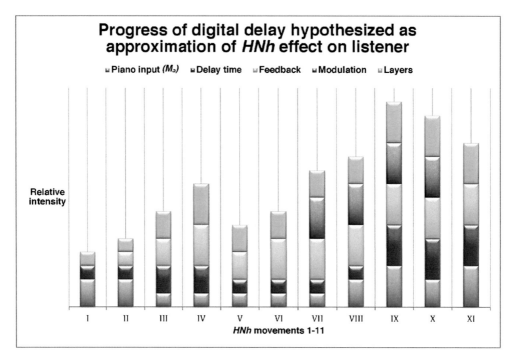

Figure 6: Simulation of a hypothetical performance of *HNh* as a manifestation of the anti-anthropocentric attractor.

This anti-anthropocentric attractor contraindicates the mental association of the term *music* with human communication. Composers (the present author included) are always trying to endow their art with a sense they imagine they can cause an audience to perceive. If a work of art is ostensibly trying to communicate something—that is, if the term *musical language*, for instance, is not taken to be an oxymoron—and so does not threaten to unravel human epistemology as currently understood, it is only because it cannot communicate anything that is not projected onto it from within the minds of composers and percipients alike; if we cannot color it with our psyches, it is truly alien. The end of the Enlightenment concept of man and the related ideology of humanism is the state to which the next stage in the evolution of art will be attracted and which it will tend to imitate.

For more about *HNh* and other works by the author, see Gillespie (2015).

7. Conclusions: heuristic value of an anti-anthropocentric compositional theory (ACompT)

An ACompT would study implausibilities in art as entities with an anti-humanistic raison d'etre. It could be based on the cybernetic feedback-loop approach and the disendowment of compositional events' anthropocentric meaning as a result of an anti-humanistic creative urge expressed by an implausibility (or implausibilities), which leads to forms of art unperceived by an audience as representations of "visions and movements" of the human mind (Brün, 2004a, p. 99). To understand the phenomena attributable to anti-anthropocentric art, such a theory would try to uncover and clarify the feedback-loop action and the necessitating creative impulses (which incite a composer to create implausible things) that maintain it. With the aid of the hypothesized anti-anthropocentric attractor, it would demonstrate the dynamics of implausibilities and their cohorts.

A question arises: Would an ACompT be able to explain many and varied phenomena if it is based on a simple modus operandi? To answer this, we might turn to the procedural technique for explaining a phenomenon (Mella, 2014), which is used whenever a phenomenon is the result of the application of some heuristic. Genetic algorithms, for example, represent a powerful procedural explanatory tool for methods of musical composition based on Darwin's theory of evolution (Fox, 2006).

If the answer is "yes," an ACompT would represent an efficient tool of system thinking aimed at providing a procedural explanation of anti-anthropocentric art, as it would determine and explain how implausibilities arise and evolve by determining and examining the rules constituting the interactive mechanisms between human creative behavior and the draw toward the implausible.

Does anti-anthropocentric art constitute a particular class of art which may be termed obfuscatory? Anti-anthropocentric art does not differ from such art and, thus, from the "ontological mutant" (Tormey, 1974, p. 204) that is music itself and much less from the disorienting ambiguities of much contemporary musical composition. First, this is because anti-anthropocentric art necessarily foments endless debates on its identity and, generally, myriad contradictory assessments of its value (Gillon, 2017) as a part of the feedback loop. Yet implausibility is not only characteristic of the experimental music of the late twentieth century, but generally of the *con*ception of music since its *in*ception (Macklin, 2010). Second, there is the irrationality and incommensurability of human perception itself:

> Perhaps, as some scientifically minded philosophers think, things really have no colours, and all our colour attributions are false. Perhaps, due to some sceptical scenario, there isn't a real, mind-independent world out there at all. Perhaps creatures from radically different cultures would conceptualize the world in ways that are inconsistent or incommensurable with our own, and there is no rational way of choosing between these rival conceptualizations. (Currie, 1995, pp. 111–112)

Third, in a manner similar (if not identical) to the absence in music of human communicational foundations, scope, and validity, the purpose of all anti-anthropocentric (obfuscatory) art remains occluded. Fourth, the anti-humanistic effects of obfuscatory art are arguably either a match for or outshine the compositional technique (or techniques) which produced them. Yet an equivalence between human creativity and compositional technique is considered to be an indispensable explanatory characteristic. Giving such importance to this equivalence has historically meant that the analysis of art must focus its attention on compositional technique in the pursuit of human understandability.

A valid ACompT would be neither a non-human approach, as it would refer to human creativity, nor a human approach, as it would also include anti-humanistic tendencies in the compositional model. It would be a human-non-human approach, or a nondualistic paradigm (Grimes, 1996), in that the governing principles describing the compositional process would have to involve the feedback between simultaneous human and non-human rules:

> At the center of blasphemous life [i.a., implausible phenomena] is this idea of the living contradiction. [Implausible phenomena] is [phenomena] that is ... but that should not be ... This contradiction is not a contradiction in terms of ... science; [implausible phenomena] can often be scientifically explained and yet remain utterly incomprehensible. If it is a logical contradiction, it would have to be one in which the existence of true contradictions would not only be admitted, but would be foundational to any ontology. (Thacker, 2011, pp. 103–104)

Because of the mutual indiscernibility of all music and obfuscatory art, neither of which can be expressed as language and so can neither be understood nor misunderstood in terms of any communicative means, a valid ACompT would represent a useful theoretical model to be applied to the study of implausibility in art. Other interpretative models may be possible; that offered by an ACompT would provide an efficient and effective interpretation of implausible phenomena considered to derive from initial compositorial inputs that are subsequently perpetuated by the feedback-loop action.

Even more interesting is the possible application of an ACompT to implausible phenomena for which we cannot immediately observe or postulate an overtly anti-human compulsion, but for which there is a fluidity and evanescence corresponding to the transcendence of the human-non-human dichotomy: Will we ever manage to define what art is? Will it ever be possible to define a work of art? Will we ever be able to say what a poem actually is? Will there ever be an end to music history? Will we live to see the end of painting? Will the posthuman continue to sculpt? to write novels? If so, how will these art forms coevolve with a rapidly evolving population? If we hope to find answers to these and many other questions, we must understand anti-anthropocentrism. If we can arrive at a valid ACompT, the answers may reveal a posthuman world and, in doing so, highlight our present ignorance, hubris, and frailty.

Three aspects of such a theory would seem to make it particularly incisive:

1. It would not just describe an instinctual, subconscious nature beyond rational control based only on general rules and thus on the analysis of individual

compositions, but would try to uncover and explain the feedback between the posthumanist paradigm and elusive human self-recognition or any necessarily arbitrary attempts at delimiting humanity or its art ontologically.

2. Notwithstanding the difficulty of defining the phenomena attributable to anti-anthropocentric art, the theory would try to uncover and clarify the link between incipient creative impulses (which cause the human behavior giving rise to yet contingent upon social, political, and cultural entities) and anti-anthropocentric factors (which work to destroy and discard the notion of the individual autonomy and sovereignty of human reason). The theory might reach the same conclusion as the present study that, in the presence of suitable creative and anti-anthropocentric factors, the self-deluding image-making of the ego will trigger the composition of a work of art, which, by *sum-ergo-cogito* nondualism, is continued and becomes, when completed, something like an instance of Lacanian *méconnaissance* (Wyatt, 2010).

3. The procedural explanation offered by the theory would not only allow us to better understand the operating mechanism producing the phenomena under investigation, but the influence epistemic assumptions of humanism have on anti-humanistic rebellion.

The challenge of an ACompT is threefold:

1. to develop more sophisticated anti-anthropocentric attractors;

2. to apply the theory to implausibilities operating in the atavistic hierarchies supposedly moving toward a state of greater and greater human civilization and to specify, for any implausible phenomenon, the creative and anti-anthropocentric factors which allow us to cognize the Other independently of delineating what humanity is not; and

3. to identify the mechanisms which, evidenced in both anti-humanism and humanism, modify the historical ideal of humankind, thereby influencing the contrarian suspicions of artists to portray the human race as a transitional experiment considered to be unfixed, narrow, transient, and fraught with danger.

References

Beer, S. (1994). *Beyond dispute: The invention of team syntegrity.* Hoboken, NJ: Wiley.
Boulez, P. (1991). Aléa. In S. Walsh, (Trans.) & P. Thévenin (Ed.), *Stocktakings from an apprenticeship* (pp. 26–38). Oxford, UK: Clarendon Press.
Brün, H. (2004a). Against plausibility [1963]. In A. Chandra (Ed.), *When music resists meaning: The major writings of Herbert Brün* (Vol. 2; pp. 89–100). Middletown, CT: Wesleyan University Press.
Brün, H. (2004b). Drawing distinctions links contradictions [1973]. In A. Chandra (Ed.), *When music resists meaning: The major writings of Herbert Brün* (Vol. 2; pp. 225–236). Middletown, CT: Wesleyan University Press.

Burke, S. (2008). *The death and return of the author: Criticism and subjectivity in Barthes, Foucault and L* Edinburgh: Edinburgh University Press.
Centre Pompidou. (1985, March 28–July 15). *Les Immatériaux*. Paris.
Currie, G. (1995). *Image and mind: Film, philosophy and cognitive science*. Cambridge, UK: Cambridge Press.
Davies, S. (2010). Perceiving melodies and perceiving musical colors. *Review of Philosophy and Psych*, 19–39.
Festinger, L. (1957). *A theory of cognitive dissonance*. Palo Alto, CA: Stanford University Press.
Fox, C. (2006). Genetic hierarchical music structures. In G. C. J. Sutcliffe & R. G. Goebel (Eds.), *Proceedings Nineteenth International Florida Artificial Intelligence Research Society Conference* (pp. 243-247). Menlo CA: American Association for Artificial Intelligence Press.
Fox, N. (2013). Creativity, anti-humanism and the 'New Sociology of Art.' *Journal of Sociology, 51*(3), pp. 522–53
Gigliotti, C. (2008). Leonardo's choice: The ethics of artists working with genetic technologies. In S. Gill (Ed), *Cognition, communication and interaction* (pp. 536–548). London: Springer.
Gillespie, Z. (2015). *A new contemporary endeavor*. http://zane-gillespie.squarespace.com/
Gillon, L. (2017). But Is It Art? In *The uses of reason in the evaluation of artworks* (pp. 15–48). Cham, Switzerland: Palgrave Macmillan.
Gleick, J. (2008). *Chaos: Making a new science*. London: Penguin.
Griffiths, P. (1981). Moments of parting. In *Modern music: The Avant Garde since 1945* (pp. 138–154). New York: George Braziller.
Grimes, J. A. (1996). *A concise dictionary of Indian philosophy: Sanskrit terms defined in English*. New York: State University of New York Press.
Harris, S. M. (2017). Deciding well. In *Boundless reason: A universal strategy for deciding well* (pp. 1–13). Naples, FL: Recursionist Publishing.
Holbrook, D. (1980). Can the artistic life survive? *PN Review 14*, 6(6), 69–70.
Holbrook, M. B. (2012), Introduction to Part IV, in *Music, Movies, Meanings, and Markets: Cinemajazzamatazz*. New York: Routledge, pp. 172–178.
Jung, C. G. (1969). *The archetypes and the collective unconscious*, (G. Adler, G. & R. F. C. Hull, Eds. & Trans.). Princeton, NJ: Princeton University Press.
Keel, J. (2002). *Our haunted planet*. Lakeville, MN: Galde Press, Inc. (Originally published in 1971)
Louis, R. (2000). Mahler. In *Lexicon of musical invective: Critical assaults on composers since Beethoven's time* (N. Slonimsky, Trans.; pp. 120–123). New York: W. W. Norton & Company.
Macklin, C. (2010). Musical irrationality in the shadow of Pythagoras. *Contemporary Music Review: Impossible Music, 29*(4), 387–393.
McWilliam, D. (2015). Beyond the Mountains of Madness: Lovecraftian cosmic horror and posthuman creationism in Ridley Scott's *Prometheus* (2012). *Journal of the Fantastic in the Arts, 26*(3), pp. 531–545.
Mella, P. (2014). The pillars of learning: Understanding, studying and explaining. *Creative Education, 5*(17),1615–1628.
Palk, A. C. (2015). The implausibility of appeals to human dignity: An investigation into the efficacy of notions of human dignity in the transhumanism debate. *South African Journal of Philosophy, 34*(1), 39–54.
Pop, M. (2014). Humanistic episteme in sustainable development of creative industries. Paper presented at the 5th International Conference on Advanced Materials and Systems. Retrieved August 6, 2019 from https://www.researchgate.net/profile/Marlena_Pop/publication/296846611_Humanistic_episteme_in_sustainable_development_of_creative_industries/links/56db259808aebe4638beeb55/Humanistic-episteme-in-sustainable-development-of-creative-industries.pdf
Prigogine, I., & Stenger, I. (1984). *Order out of chaos: Man's new dialogue with nature*. Boulder, CO: Shambhala Publications.
Russell, B. (1961). On induction [1912]. In R. E. Egner & L. E. Denonn (Eds), *The basic writings of Bertrand Russell, 1903-1959* (pp. 121–127). Crows Nest, AU: Allen & Unwin.
Scott, H. E. (2009)."Wholly uninteresting": The motives behind acts of iconoclasm. In *Confronting nightmares: Responding to iconoclasm in Western museums and art galleries* (pp. 39–49). Thesis, St Andrews Research Repository, University of St Andrews, Fife, Scotland. Retrieved from http://hdl.handle.net/10023/788 on July 2, 2019.
Tormey, D. (1974). Indeterminancy and identity in art. *The Monist, 58*(2), 203–215.
Thacker, E. (2011). Nine *disputatio* on the horror of theology. In *In the dust of this planet: Horror of philosophy, Vol. 1* (pp 98–132). Alresford, UK: Zero Books.
Virilio, P. (2003). *Art and fear*. London: Continuum.
Wyatt, J. (2010). Mirror stage of identity development. In R. L. Jackson & M. A. Hogg (Eds), *Encyclopedia of identity, Vol. 1* (p. 463). Thousand Oaks, CA: Sage.